# Biochemistry
## Lab Manual

Kelli Slunt

Kendall Hunt
publishing company

# Table of Contents Biochemistry Laboratory:

# Practices and Techniques Used in the Biochemistry Laboratory

Biochemistry is an exciting discipline that combines the chemical understanding of structures and functions of molecules with the biological study of life. This increasingly changing field assists in scientific advances in related areas, such as forensics, the pharmaceutical industry, and genetic engineering. This laboratory manual will provide you with an introduction to, and background for, many techniques utilized in this advanced field of study.

Many of you are preparing for careers in medicine, the pharmaceutical industry, academic research, and government. In order to be employable and successful in a scientific career, strong critical thinking skills and good laboratory practices must be developed. Scientists must not only be able to collect data but, more importantly, interpret their findings. This requires the ability to think independently and critically about scientific results. Since a successful biochemist must first understand the theory and execution of basic techniques, especially biochemical analytical skills, each chapter of this text will provide background and application of many commonly used biochemical methods.

Good laboratory practices are standards that are followed in industry and academia to ensure that data and laboratory results are collected and recorded in a safe, accurate, and reproducible manner. Historically, unethical practices or gross scientific errors have led to instances of distrust of scientists and/or detrimental scientific setbacks. (See Box 1.1.) Following good laboratory practices is essential to your success in this course as well as in future endeavors.

---

### Science in Action

#### *Box 1.1 – Scientific Fraud*

Scientific fraud, although rare, does occur, and a number of cases have been publicly exposed in recent history. In most scientific fraud cases, the scientists misrepresent or misreport data in order to support proposed theories. Fewer research fraud cases involve plagiarism or data falsification. In August 1999, scientists at Lawrence Berkeley National Laboratory (LBNL) published results in *Physical Review Letters* indicating that they had produced and detected atoms of elements 116 and 118. In 2001, it was discovered that the data were fabricated, and, as a result, nuclear scientist Victor Ninov was fired from his position at LBNL. In another case, Jan Hendrik Schön was fired from Bell Labs because he reported fraudulent or substituted data at least 16 times over a three-year period. (*C & E News,* 2002, **80**, 44, pp. 31-33). Both of these cases were uncovered because other scientists could not reproduce their results. The United States federal government is tightening regulations to ensure that scientific data are not fraudulent. In 2001, the government fined a Linden, NJ, laboratory that had falsified laboratory reports to show that a gasoline product met Clean Air Act standards. (*C & E News,* 2002, **80**, 13, pp. 49-50).

In addition to ensuring credibility, proper laboratory practices reiterate the case for scientists to act safely and responsibly. Biochemistry is an experimental science that involves the use of equipment and chemicals that are potentially hazardous. The safety and waste management points discussed in detail throughout this manual should be closely followed.

## 1.1 PREPARING FOR THE LABORATORY

The laboratory can be a safe environment to discover fundamentals about science; however, if not respected, the lab can also be a place of danger and accidents. One of the best ways to have a safe and stimulating laboratory experience is to arrive well prepared. Prior to each experiment, you should read through the introductory material and the experimental procedure to be followed. If necessary, read or research relevant background material in textbooks, literature, or reference materials (see Searching/Using the Scientific Literature, below). Your goal is to arrive in the laboratory with a plan in mind, a proposed strategy to execute the experiments in a safe and efficient manner.

## 1.2 SEARCHING/USING THE SCIENTIFIC LITERATURE

The scientific literature is a very important tool for any scientist. Prior to the start of a new project, a scientist will consult the literature to determine previous work conducted in the area. The literature often provides information about experimental techniques, background, or the current state of knowledge. A scientist also routinely reads the literature to determine the significant unknowns in his or her field of study. This yields novel ideas for research. In order to develop into a practicing scientist, you should know how to use the literature effectively and efficiently.

There are several components to the biochemical literature. These will be discussed in more detail. Primary literature sources are those where most of the information presented is being reported for the first time. The authors of primary literature, particularly in the sciences, report firsthand on their research findings. Examples of primary sources include some articles in journals (usually peer-reviewed scholarly journals), patents, papers in conference proceedings, some dissertations and theses, some technical reports, and laboratory and field notes. Secondary sources provide review and summary of information from the primary literature, as well as references to the primary literature. Secondary literature sources include reviews, some books, and reference materials, such as encyclopedias, book series, and databases. Students may want to begin with secondary literature, as it provides a place in which ideas are integrated and reviewed. New findings/data and details are generally not found in the secondary literature, and reference to the primary literature would be required.

The following list outlines important components of the scientific literature (both primary and secondary sources):

1) **Textbooks** – Textbooks offer a starting basis for biochemical information. Unfortunately, the information found in textbooks is at least 1 to 2 years old by the date of publication. Textbooks, however, do provide a good summary of the findings and a survey of the topic. Also, many textbooks list references to research or review journals that are pertinent to the topic.

2) **Reference books and review publications** – Reference books are good reviews of recent topics in biochemistry, often of a more specific subject, such as enzymes or cytochrome P450. Typically, reviews are published annually or semiannually and summarize current

research in an area. Examples include *Annual Reviews of Biochemistry* and *Trends in the Biochemical Sciences (TIBS)*.

3) **Research journals** – This is probably the most important resource available to an experimentalist. Refereed or peer-reviewed journals are more respected than other journal publications, as the articles are critically reviewed by experts prior to publication. Research journals contain the most up-to-date advances in science. Journals are available in print, on microfiche or microfilm, and also electronically. Several examples of research journals are *Journal of Biological Chemistry, Biochemistry, Science, Nature,* and *Proceedings of the National Academy of Sciences.*

4) **Methodology references** – Several research journals and publications specialize in providing information about new methods for experiments. These journals include *Analytical Biochemistry, Analytical Chemistry, BioTechniques,* and *Methods in Enzymology.*

5) **Handbooks of chemical and biochemical data** – These works supply information about chemical and biochemical substances (i.e., physical and chemical properties, molecular weights, etc.) Examples include the *CRC Handbook* and *Merck Index.*

6) **Computer-based search engines** – In order to find information in the literature, one generally starts with a literature search. Common places to begin include Web-based search engines or *Chemical Abstracts*. There are numerous search engines and databases available, including *SciFinder, Beilstein, Current Contents, PubMed,* and *Science Citation Index.* Protein and nucleotide databases can be searched using *BLAST.*

7) **Molecular modeling resources** – The visualization of biochemical molecules is extremely important. Several programs, such as RasMol, Kinemage, and Protein Explorer, have been written to aid researchers in viewing these molecules. Data files can be downloaded for all known published structures from databases such as Research Collaboratory for Structural Bioinformatics (RCSB), Protein Data Bank, Brookhaven Protein Data Base, and Genbank.

# 1.3 SAFE LABORATORY PRACTICES

After experimental plans are developed, scientists enter the laboratory to execute their plans. Laboratory accidents can occur at any time, even with the most cautious person. Therefore, it is important that you use common sense and chemical safety procedures. Most laboratory accidents are caused by the incorrect handling of chemicals, exposure to chemicals, burns associated with fires or explosions, or cuts from broken glassware. Some accidents, however, can occur due to carelessness: for instance, a drawer left open or water spilled on the floor (from washing glassware).

The following list can serve as guidelines for safe practices in the laboratory:

1) As undergraduate students, you should never work in the laboratory alone. Do not conduct any unauthorized experiments. In case of an emergency or laboratory accident, contact the instructor or another person immediately.

2) Wear appropriate eye covering, such as goggles, for all laboratory work. If you should get a chemical in your eye, immediately wash with flowing water for 15-20 minutes, have someone inform the instructor, and seek appropriate medical attention after rinsing the eye.

3)  Do not eat, drink, or smoke in a laboratory. Do not bring food or drink into the laboratory, and do not taste any of the laboratory chemicals (including ice). Do not eat or drink from laboratory glassware. Additionally, the insertion or removal of contact lenses or the application of cosmetics or lip-balm should be conducted outside the laboratory to prevent the transfer of chemicals to your face or skin.

4)  Wear appropriate clothing in the laboratory. Sandals, shorts, and/or unconfined long hair can pose a safety hazard. Also, certain chemicals can discolor or damage some fabrics. A laboratory coat or apron will protect your clothing and exposed skin. The appropriate gloves should be worn to protect your skin from hazardous chemicals or materials.

5)  Exercise care when detecting the odor of fumes. Avoid breathing fumes of any kind. If a reaction or bottled reagent can give off toxic fumes, work with these items in a fume hood.

6)  Do not use mouth suction to fill pipettes or start siphons.

7)  Know the location and use of all safety equipment in the laboratory.

8)  Work in an uncluttered area. Store all equipment and chemicals in their proper places.

9)  Avoid contamination of laboratory chemicals. Do not return any excess material to its original container. If a small amount of reagent is transferred to another container, make sure the smaller container is well labeled. Do not have beakers or flasks with unlabeled substances in the laboratory.

10)  Dispose of chemicals using appropriate protocols.

11)  Be familiar with the properties and hazards of all chemicals in the laboratory. Information about the reactivity, flammability, toxicity, and possible hazards of chemicals can be found on Material Safety Data Sheets (MSDS). Information about MSDS can be found on-line at numerous sites, including www.msdsonline.com, www.ilpi.com/msds/index.html, and chemical manufacturer websites.

## 1.4 RECORDING DATA

Academic and industrial scientists carefully document ideas, results, and observations from experiments. Scientists accurately and attentively maintain a laboratory notebook. This serves as a record of specific documentation: for example, to prove date of conception of an idea or finding (necessary for patent applications or proper recognition of ideas). In order to prepare you for a future career as a scientist, good record-keeping skills must be developed.

A laboratory notebook can be viewed as a working document or a diary of scientific activity. It can serve as a chronological account of your work. All records of your work should be recorded neatly and readably in a bound notebook with consecutively numbered pages. It is imperative that you keep detailed notes on your experiments. No absolute format for a notebook exists, but the guidelines listed below should be followed.

- Write all entries legibly in permanent black ink.
- Date each entry in the notebook. (Industrial employers require each page to be signed and dated.)

- Make entries in chronological order.
- Create a table of contents at the beginning of the notebook. The table of contents should contain a brief title/description of each experiment and the pages in the notebook where the experiment can be found. This will enable you to find a piece of earlier data without hunting through pages, searching stacks of data, or looking in a drawer.
- Use consecutive pages. Leaving blank pages is unethical, as it implies that you plan to return at a later date to add data or ideas. (See Box 1.1.) If necessary, you can put a reference at the bottom of the page to a future reference in the notebook (e.g., on page 10, you could write, "See page 35."), or draw a diagonal line through the skipped portions of the notebook.
- Do not erase entries. If you feel you made a mistake, cross through the error with a single line. Do not scribble out the entry or use Wite Out®. You may find later that it was not a mistake and the information is important.
- Do not remove pages from the notebook.
- Record the procedure followed, any notes, and all experimental observations and data.
- Provide enough detail that another person could successfully repeat your work.
- Label all figures, graphs, tables, and calculations.
- Supplemental material may be added to your notebook as long as it is permanently affixed to the notebook in its proper chronological location.
- Write directly in your notebook while you are in the laboratory and performing experiments.
- Do not rely on notes on scraps of paper. Do not rely on your memory.
- Show all calculations in your notebook.
- Include the date when comments or annotations are added after the fact.
- If you have an idea that is patentable, have the pages of the notebook witnessed and signed by a peer.

As mentioned above, there is no published format for a laboratory notebook; however, certain standard sections will make a notebook easier to read. You may want to include the following in the notebook entries related to each experiment:

- Title or descriptive heading for the experiment.
- Objectives for the experiment. These describe the purpose or goal(s) of the experiment and the hypothesis.
- Experimental procedure or materials/methods. This section is a record of the protocols, methods, materials, and solutions used in the experiment. You should provide enough detail that another person could reproduce and verify the experiment. If the experimental procedure followed comes from a published source, it is appropriate for you to firmly attach a copy of the procedure to the notebook or to cite the reference. Any modifications to the procedure should be noted in detail in the notebook.
- Results. This section is a record of all observations, data, graphs, photographs, etc. The data must be legible, well organized, completely labeled, and written in a fashion that could be understood by someone else and by yourself (especially at a later date). Note what you have done and what you have observed. In addition, one could note what you should have done and what you think should have happened, but do not do this exclusively.
- Conclusions/data analysis. This section is a short analysis of the findings. The analysis may be graphical or tabular, as long as there is written text explaining the relevant points of your figure or table. A summary chart may simplify this section.
- Comments. This section enables you to describe any possible experiment errors, suggest improvements in the protocol, etc. Comments should also be written throughout the

course of the lab, noting, for example, any observations, abnormalities, or baffling results. Comments or annotations added to your notes at a later date should include the date on which they were added.

## 1.5 WRITING GOOD LABORATORY REPORTS

A laboratory notebook provides a chronological record of the progress of experiments as they are conducted, primarily for the benefit of the author or other scientists in the laboratory. Scientists use laboratory reports or scientific papers to communicate their experimental findings to a wider audience in the scientific community or within a corporation/organization. A laboratory report should be written in a format similar to scientific papers in the literature. The document must be written coherently at a level for someone who has some experience in the field (i.e., another biochemistry student or a colleague). Generally, reports are written in the past tense, third person, and passive voice. Although the exact titles or sections, referencing styles, and report formats may differ slightly, most laboratory reports contain the following sections:

1) **Abstract** (typically 100 words or less) – This is a brief summary of the entire report. The abstract should include a statement of the purpose of the experiment, the general method used, the result(s), and the major conclusion(s). You may find it easier to write this section last. Some students incorrectly write this section as an introduction. It is not an introduction, but rather a summary. You should be able to read and understand the abstract and the paper independently from one another.

2) **Introduction** – In this section, you should provide the minimum necessary background material to understand the experiment. It should be written for someone who has some experience with the field but is not familiar with the experiment (i.e., another biochemistry student who has not taken the laboratory course or someone working in a similar laboratory). There should be references to the existing literature in this section. Also in this section (usually near the end of the section), you should present the hypothesis and rationale behind the experiment.

3) **Experimental** (Materials and methods) – This section contains a careful description of the procedures, protocols, materials, and solutions used in the experiment. The information should be sufficiently detailed for the informed reader to understand and duplicate your experiment. Statistical methods used to analyze the data should also be reported. If the procedures used are adequately described in a previous publication, for example in a journal article or lab handout, it is generally sufficient to describe them briefly and reference the source. Any changes or additions that have been made, however, must be clearly noted. The quantities of materials used and their units should be specified here. The manufacturer of instrumentation or chemicals should be noted in this section.

4) **Results** – In this section, you should present the findings of the experiment, in narrative form, supplemented, when appropriate, with figures, tables, diagrams, graphs, illustrations, etc. Each figure or table should be presented on its own page at the appropriate point in the paper or inserted at the appropriate location in the text. The figure or table should be numbered appropriately, have a descriptive title, and contain enough information (e.g., reference citation, legends, captions, etc.) to make it understandable. You may want to refer to journal articles for examples of figures and tables in the literature. You should refer to your figure or table in the text and point out the important features rele-

vant to the hypothesis of the experiment. The figures and accompanying caption/text in the results section alone are not enough; you must include text. You do not have to write out all of the results in the text, but you should describe the most important findings. The most difficult part of this section is to report the results without interpreting them in the context of your hypothesis, experimental objectives, or the literature. The discussion of results comes in the next section of the paper. If you prefer, the results and discussion sections may be combined.

5) **Discussion** – This section is the most important portion of the manuscript; it contains a detailed analysis and interpretation of the findings reported in the results section. Here, you must evaluate your hypothesis in light of the experimental data. You should compare the data obtained in your experiment to that reported in the literature. You should also comment on sources of error and discuss future possible studies. A description of the relevance of the results to larger questions in the field or to related fields may be appropriate as well. You should not restate the results in the discussion section, but you should demonstrate that you have thought about the meaning and relevance of your findings.

6) **Acknowledgements** – In this section, you should acknowledge any assistance you received from laboratory partners, resources such as a writing center, a collaborator, the instructor, etc. If the research was supported by a financial grant, it is appropriate to acknowledge the support in this section.

7) **References** – This section should include a complete bibliography of the literature cited in the text. There are several different and acceptable ways to cite the literature in chemical and biochemical journals. You should refer to *The ACS Style Guide* for assistance in citing the literature. (See References and Further Readings.)

## 1.6 UNCERTAINTY/ERROR ANALYSIS

Each measurement made in the laboratory carries a degree of uncertainty or error. The amount of uncertainty often depends on the accuracy of the measuring device or instrument. The last digit in a measurement is the place where the uncertainty lies. For example, if one measured 5 mL with a 100 mL graduated cylinder, the uncertainty or error is ±0.5 mL. In order to obtain a more accurate measurement, a more accurate device needs to be used, such as a smaller graduated cylinder, a buret, or pipet. Errors associated with a particular measuring device are often referred to as "systematic errors." These are often very significant to the collection of experimental data and can be caused by instrumental calibration errors, maladjustments, or environmental factors. For example, the use of a metal ruler to measure objects depends on the temperature of the environment. At colder temperatures, the metal contracts and the distance between calibration marks on the ruler becomes smaller than at a significantly higher temperature. Temperature also impacts volumetric and gravimetric measurements. Glassware is calibrated for use at specific temperatures, and the calibration is printed on the glassware.

### *Random and Human Errors*

Another type of error is random error. Random errors are often less predictable and harder to control than systematic ones. Examples of random errors include situations where the observer must make a judgment decision about a measurement or unpredicted changes in environmental conditions. Random errors frequently occur because measurements require judgments about perception, and humans cannot perceive anything with exactness. Finally, human errors, such as incorrectly reading or recording a measurement or adding incorrect amounts of reagents, can also occur in experiments. An example of random versus human error follows:

> Several individuals are measuring the length of an object using a ruler. One estimates the length to be 244.56 cm, but records the length as 244.65 cm. This is an example of human error. Another person observes that the measured length is 244.58 cm, while his lab partner observes the length as 244.55 cm. This difference of 0.03 cm is a random error and is due to different perceptions by the human observers.

## *Statistical Treatment of Experimental Data*

### The Mean

In order to minimize the error associated with measurements, scientists collect more than one set of data. It makes sense that a more accurate result is obtained if the measurement is made 10 times instead of one. But after you obtain a set of data, how are the data analyzed and reported? First, the average, or arithmetic mean, of the data is determined. The mean of the set of data is equal to the sum of all of the numbers in the set divided by the number of values in the set:

$$\mu = (x_1 + x_2 + x_3 + \ldots) / N \text{ or } (1/N) \sum_{n=1}^{N} x_n \, ,$$

$$\text{where} \qquad \mu = \text{mean}$$
$$x_n = \text{individual data points in the set}$$
$$N = \text{number of data points in the set or sample size}$$

Generally, the mean provides the best estimate of the true value for the measurement. For normal distributions of data, the mean is the most efficient and least variable of statistic calculations.

### The Median

The median is the middle data point in the distribution. Half of the data points are above the median, and half the data points are below the median. The median is a better statistical indicator than the mean for skewed or non-symmetrical distributions of data.

### The Variance

The variance is a measure of the spread of the distribution of the data. The variance ($\delta$) for large sample sizes is calculated as the average of the squared deviation of each number from the mean of the data set.

$$\delta = (1/N) \sum_{n=1}^{N} (x_n - \mu)^2$$

For sample sizes of 20 or less,

$$\delta = (1/[N-1]) \sum_{n=1}^{N} (x_n - \mu)^2$$

### Standard Deviation

The standard deviation of the data is the square root of the variance. It is the most commonly used indicator of spread. More advanced statistical treatment for experimental data, such as t-tests or Chi square tests, may also be necessary. Further information about these tests can be found in books about statistics or on websites such as www.davidmlane.com/hyperstat/index.html.

## 1.7 GRAPHICAL ANALYSIS OF EXPERIMENTAL DATA

Scientists often graph data to visualize the relationship between the experimental variables. Some of your experiments may require you to construct a table or a graph of your data. Whether you prepare a hand-drawn graph or use a computer program to generate a plot of your data, there are several key items to keep in mind:

1) Plot the independent variable on the x-axis. The independent variable is the experimental value that is changed by the experimenter. The dependent variable is the measured value that depends on the independent variable. Plot the dependent variable on the y-axis.

2) Select an appropriate scale for each axis on the graph. Do not crowd the data or use only a portion of the plot. The "start" for the plot depends on how the graph is to be used; thus, the start is not necessarily always zero. The starting point, however, is clearly labeled.

In the examples below (see Figures 1.1–1.4), the plot in Figure 1.1 has inappropriate scales on both axes, whereas the plots in Figures 1.2 and 1.3 are incorrect on one of the axes. Figure 1.4 shows a plot that has an appropriate scale, so that the maximum amount of the graph is utilized.

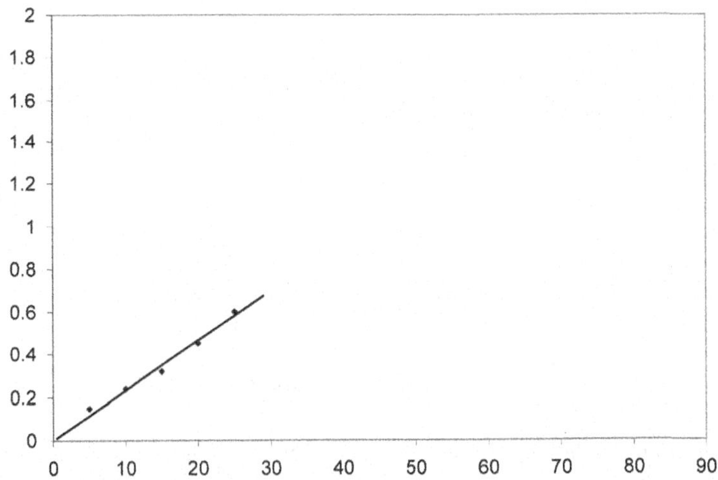

Figure 1.1 – Plot of experimental data with inappropriate scales on both axes.

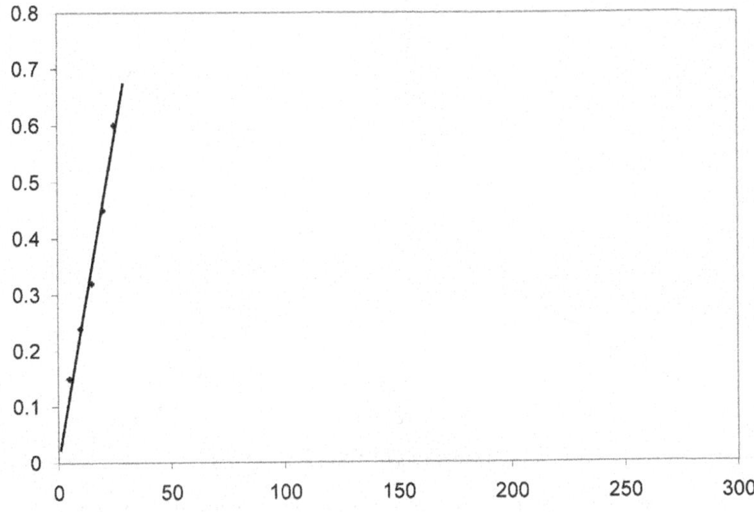

Figure 1.2 – Plot of experimental data with inappropriate scale on the x-axis.

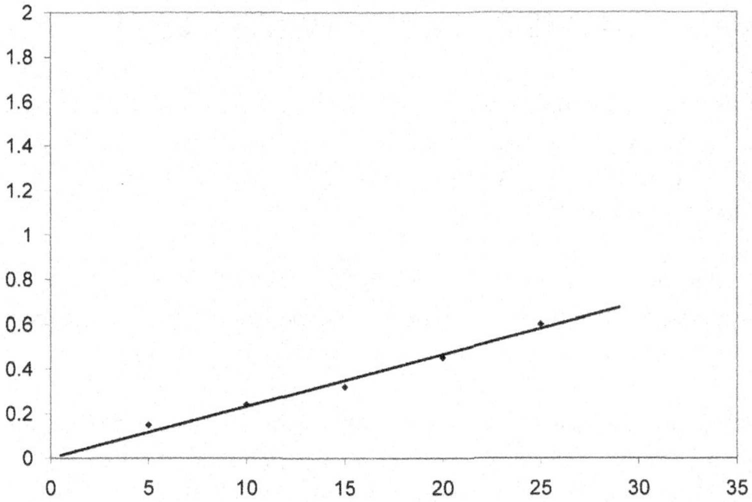

Figure 1.3 – Plot of experimental data with inappropriate scale on the y-axis.

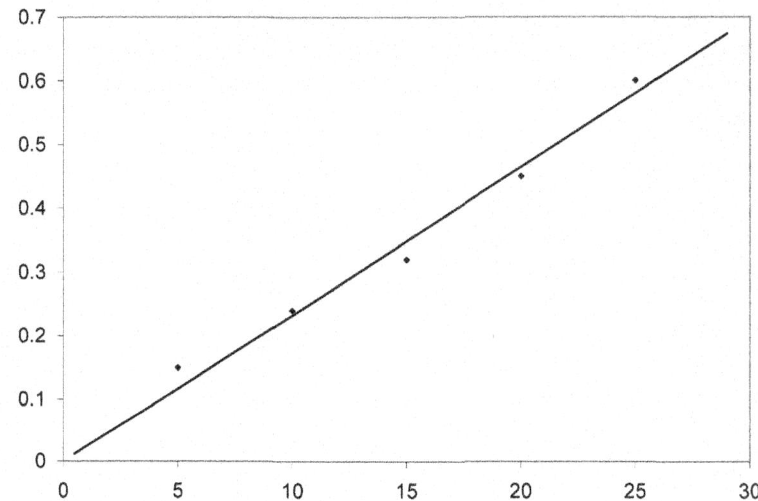

Figure 1.4 – Plot of experimental data with appropriate scales on both axes.

3) Label all axes and provide a descriptive title for the graph. Label the major divisions on each axis with numbers. Label each axis with a description of the measurements it represents. Units are usually included in parentheses. (See Figure 1.5.)

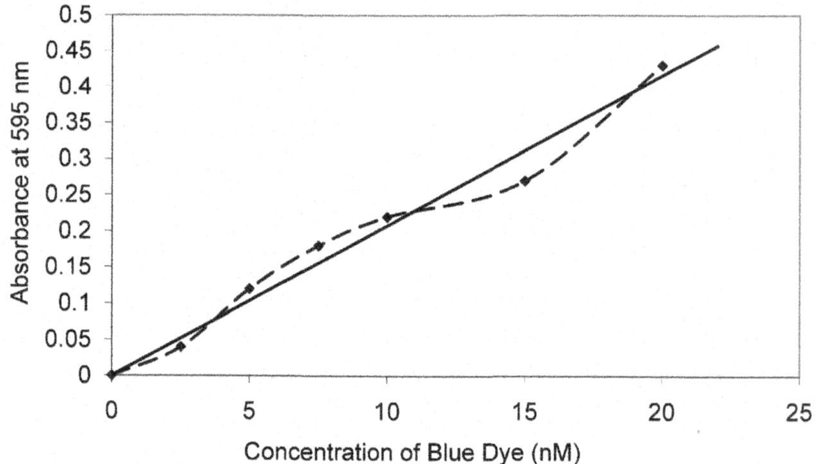


Figure 1.5 – Plot of concentration of dye versus absorbance.
Note the linear trend line better represents the data than the dashed, dotted line.

4) Do not play "connect the dots." Draw the line that best fits the data; this is often a straight line. (See the incorrect dashed line versus the solid line in Figure 1.5.)

5) If the plot is a straight line, the equation of the line is given by y = mx + b, where m is the slope and b is the y-intercept. Calculate the slope using convenient points *on the line,* not the data points. Give the equation for your graph. (See Figure 1.6.)

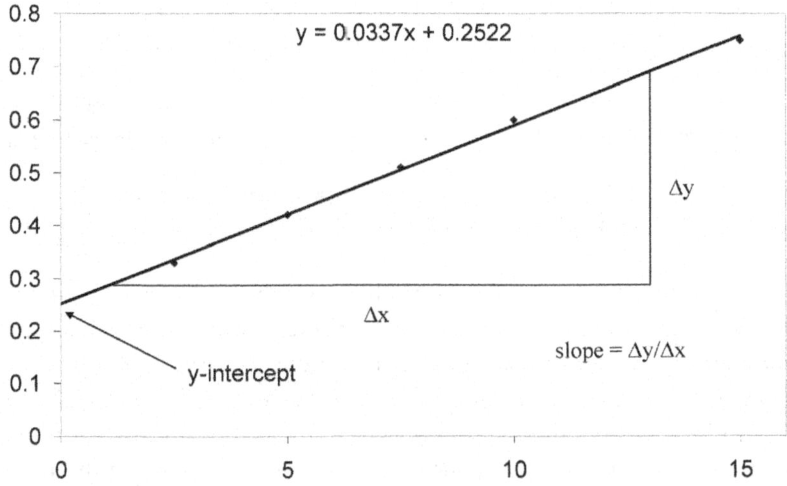


Figure 1.6 – Linear regression plot of experimental data.

6) If more than one set of data is plotted on a single graph, provide a legend that identifies what each line represents. (See Figure 1.7.)

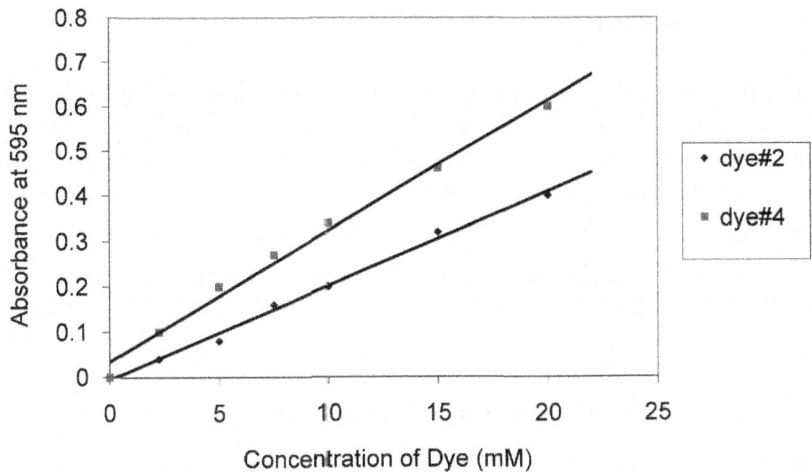


Figure 1.7 – Graph containing two sets of data with a legend indicating each data set.

7) One key to graphing data is to think about what you expect to observe. Below are two graphs of the data from a sample experiment. The difference between the two graphs in Figure 1.8 is the curve fitting.

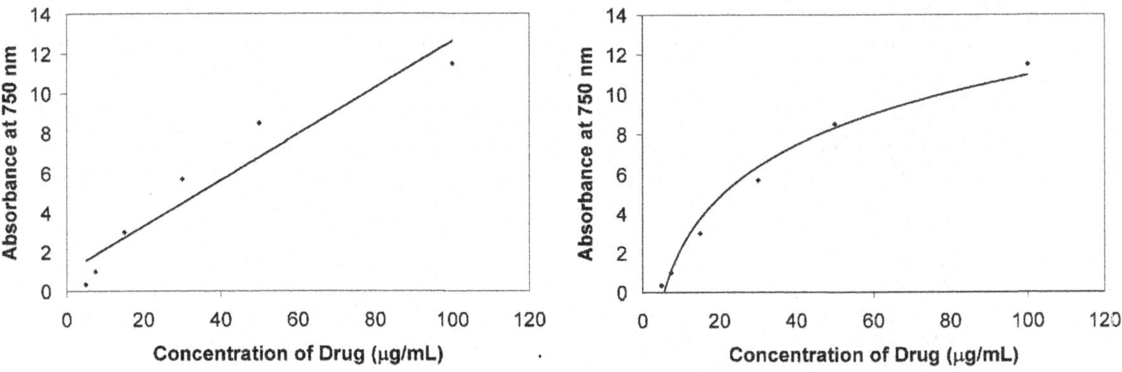

Figure 1.8 – Graphical analysis of experimental data using
linear relationship and logarithmic regression.

You need to think carefully about the relationship between the variables in your experiment and what type of graphical relationship they may have. Consider a study in which a known protein is treated with an unknown drug, and the drug-protein complex absorbs light at 750 nm. The graph on the left assumes that as the concentration of the unknown drug increases, the binding to the protein A would also increase in a linear fashion—meaning as the concentration of drug increases, the amount bound to the protein would increase proportionally. The graph on the right shows a logarithmic relationship in which there are some limitations to the binding. There is still an increase in binding with an increase in drug concentration, but the relationship is not linear. In this case, after thinking about the situation (a biological molecule that could be affected by other variables such as structural effects), the graph on the right is a better representation of the experimental results. If possible, one might want to collect more data points to fully elucidate the relationship between the variables.

The final point to keep in mind is experimental error. The curve on the left may be a linear relationship except that one point is associated with a large experimental error. A reexamination of the experimental methods and notebook may reveal that the amount of drug was measured incorrectly (or some other error occurred), and that data point may be eliminated. (See Statistical Treatments of Experimental Data.) Again, you need to think about the results and what they mean before immediately discarding a data point. If possible, scientists perform experiments multiple times so they can compare several trials to ensure that a data point is valid.

## 1.8 STANDARD LABORATORY GLASSWARE

Most glassware and equipment that you will use in a biochemistry laboratory is the same as in other chemistry courses. You should already know how to use standard laboratory glassware (e.g., beakers, Erlenmeyer flasks, volumetric flasks, graduated cylinders, etc.). A few pieces of equipment may be new (such as micropipetters), and they will be pictured and explained in this chapter. Also, this chapter will review some of the basics of measuring and transferring liquids and massing objects. Specific equipment and instrumentation that will be used in the experiments will be explained in more detail in future chapters.

It is imperative in a biochemical laboratory that you have an organized and clean workplace and equipment. Many biochemical experiments utilize small quantities of materials (e.g., $\mu$g or mg), and any contamination from the walls or sides of a glass container or pipette can drastically influence the results of the experiment. Also, many biological molecules are sensitive to metal ions, small molecules, detergents, and organic residues. These contaminants can inhibit enzymes or alter the structures of proteins, thereby altering your results. It is therefore crucial that you clean your glassware properly.

Unless otherwise noted in a procedure, wash all glassware, spatulas, etc., with a mild detergent. Rinse the items thoroughly with tap water, followed by three washes with deionized (DI) water. If you are preparing a solution with ultrapure water, follow the rinses with DI water with three rinses with ultrapure water. If you require more specialized information about cleaning using chemical solvents, ultrasonic cleaners, or base baths, please consult your instructor or a more specialized laboratory handbook. (See Coyne.)

If the procedure calls for completely dried glassware, the inside of the glassware can be rinsed with acetone, and the items placed in a laboratory oven around 100°C for 15-20 minutes. If residual organic solvent is undesirable or if the item is plastic, avoid rinsing with acetone. Plastic items should be dried at room temperature, as heating in an oven can cause distortions that could impact accuracy.

## 1.9 MEASURING AND TRANSFERRING LIQUIDS

Precision volumetric glassware is used to measure and/or deliver exact volumetric quantities of liquids. The capacity of the glassware (1 mL, 10 mL, 1000 mL, etc.) is inscribed on the device. The glassware has one or multiple calibration marks indicating exact volumes. Erlenmeyer flasks, beakers, disposable centrifuge tubes, sample vials, etc., are not volumetric glassware. The volume indicated on these glassware items indicates approximate capacity or volumes. These containers are good for storing and mixing liquids, not for measuring volumes.

Glassware, including calibrated pipets, burets, and syringes, bear the inscription TD, which means "to deliver." These items are designed to measure and deliver a particular volume of liquid. Graduated cylinders are designed to measure and deliver particular volumes of liquids but are less accurate than the other TD glassware listed. In practice, graduated cylinders are used as TD vessels for volumes of 1 mL or more. The best way to deliver a small volume of liquid is by using a pipet, buret, or syringe.

Vessels marked with a TC label are intended "to contain" a specific volume of liquid. Volumetric flasks are an example of TC glassware that are used to prepare solutions with specific volumes. Due to their unique shape, not all of the contents are often delivered, thus the designation "to contain."

The following includes steps to consider when selecting glassware for measuring liquids:

### *Guidelines for Selecting Glassware for Measuring Liquids*

1) Always select the smallest device that will accommodate the desired volume of liquid. For example, do not measure 10 mL using a 1 L graduated cylinder. A 10 mL cylinder would be a better choice. The correct device will minimize the error associated with the measurement.

2) As mentioned above, use a pipet, syringe, or buret to measure and deliver volumes under 1 mL.

3) Never use a piece of nonvolumetric glassware for accurate measurements. These devices are meant for storing and mixing.

4) Oily and viscous liquids, such as glycerol or mineral oil, are difficult to remove from certain glassware items. They require long drainage times. Consider using a disposable syringe or measuring the necessary quantity by weight instead of volume.

## *Pipets*

There are several types of pipets that are frequently utilized in a laboratory. (See Figure 1.9.) Volumetric pipets are calibrated to deliver a specified volume if the level of the liquid is aligned with the calibration line (single line about the expanded part of the glass near the middle of the pipet). Measuring pipets are calibrated to deliver various volumes of liquid. Serological (or blowout) pipets are graduated all the way to the tip. Mohr pipets are not graduated to the tip, and the volume must be stopped at an appropriate graduation line in order to accurately deliver the desire volume. Serological and Mohr pipets have multiple calibration marks. The volume delivered is the difference between the initial and final levels of the liquid in the pipet.

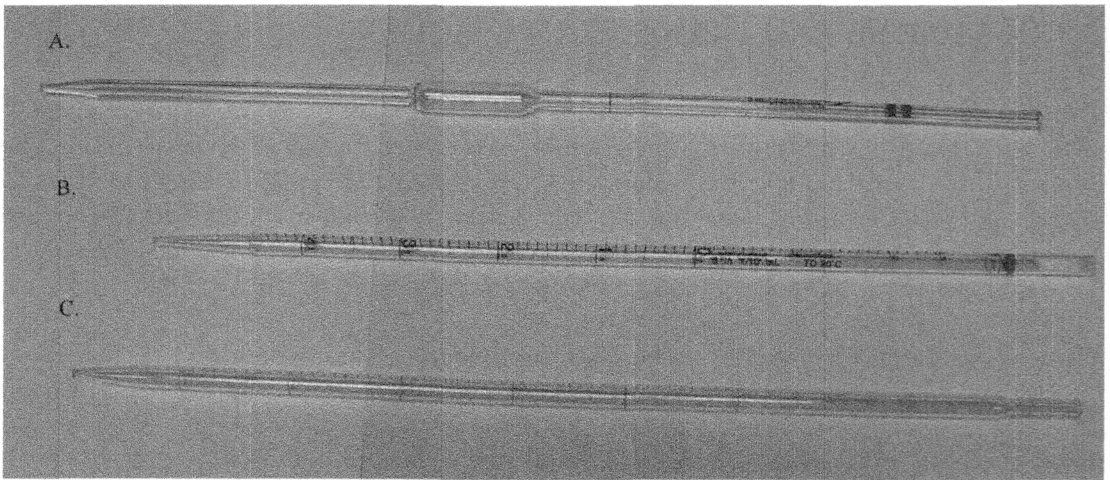

Figure 1.9 – Representative pipets: A) volumetric pipet, B) serological pipet, C) Mohr pipet.

There are general procedures that apply to all pipets. The following list details a general procedure for the use of a pipet:

1) Attach a suction bulb to the end of the pipet. Do not attempt to mouth-pipet liquids.

2) Place the tip of the pipet into the liquid, and draw the liquid level above the calibration mark on the pipet.

3) Remove the pipet tip from the liquid and slowly allow the liquid in the pipet to drain until the bottom of the meniscus is aligned with the calibration line; always view this at eye level to avoid parallax errors. (See Errors in Liquid Measurements.) Any drops clinging to the sides of the pipet can be removed by gently touching the tip to the side of the container from which you are dispensing or by wiping with a Kim wipe.

4) Transfer the pipet to the receiving vessel. Drain the liquid from the pipet. While draining, hold the pipet in a vertical position.

Volumetric pipets are designed to deliver the designated volume. Any amount remaining in the tip should be allowed to remain, as it was not part of the intended deliverable volume. Mohr pipets should be drained to the desired calibration mark, and any drops of liquid clinging to the end of the pipet should be touched to a wet surface of the receiving vessel. Serological pipets should be drained until the liquid flow ceases. Any liquid remaining in the tip should be blown out and added to the receiving container. Volumetric pipets and Mohr pipets are calibrated for the touch-off method and should never be blown out as described for serological pipets.

### *Micropipetters*

Many biological applications require the dispensing of very small quantities of liquids (i.e., micro-liters quantities). A number of commercial micropipetters are available, including Pipetman™ by Rainin and Eppendorf pipettes. Micropipetters are manufactured as adjustable or fixed volumes and can be used to deliver a single volume (single channel) or simultaneously deliver the same volume to individual vessels (multi-channel). Figure 1.10 shows an example of a Pipetman™ adjustable volume pipette. The discussion in this chapter will generally concern single channel adjustable volume pipets. If you have any questions about using a particular pipet, consult the literature provided by the manufacturer or your instructor.

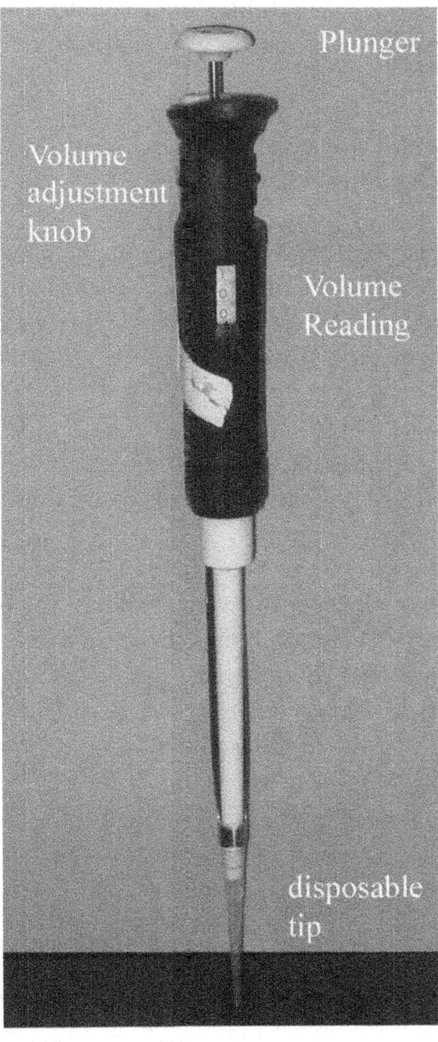


Figure 1.10 – Micropipetter (Pipetman™) with most important parts labeled.
See text for explanation of use of the micropipetter.

Most micropipetters contain an air-cushioned piston that dispenses liquid through displacement of air. Adjustable volume micropipetters come in various sizes, as shown in Table 1.1. The table shows the range over which the volume can be adjusted and the typical range that is recommended for use.

Table 1.1 – Volume Ranges for Typical Micropipetters

| Typical Adjustable Volume Range ($\mu$L) | Recommended Volume Range ($\mu$L) |
| --- | --- |
| 0 – 2 | 0.1 – 2 |
| 0 – 10 | 0.5 – 10 |
| 0 – 20 | 2 – 20 |
| 0 – 100 | 10 – 100 |
| 0 – 200 | 50 – 200 |
| 0 – 1000 | 100 – 1000 |
| 0 – 5000 | 500 – 5000 |

Micropipetters use disposable plastic tips that fit on the end of the shaft surrounding the chamber with the piston. It is important to use the proper size and type tips specified by the manufacturer. Incorrectly fitting tips can lead to leaks or inaccurate volume delivery. Tips should be stored in closed containers and you should avoid touching the narrow end of the tip to prevent contamination.

Micropipetters are calibrated for use with aqueous solutions and the following procedure should be used for forward pipetting of liquids:

1) Select the appropriate micropipetter for your measurement. (See Table 1.1 for reference.) Using the volume adjustment knob, turn slowly until the indicator shows the desired volume. Always approach the desired setting from a higher volume to prevent mechanical errors that could affect accuracy. If you pass the desired setting, dial back up and then reset the volume.

2) Attach a new disposable pipet tip on the end of the pipetter. Make sure an airtight seal is created by pressing the tip gently onto the shaft. Avoid touching the narrow end of the tip.

3) Depress the plunger to the **first** positive stop. Holding the pipetter vertically or nearly vertical (within 20 degrees of vertical), immerse the tip a few millimeters into the liquid. Allow the plunger to slowly rise to its starting position. Liquid should rise in the tip. If you are drawing air into the tip, depress the plunger to remove the liquid/air and begin this step again, ensuring that the tip is immersed into the liquid.

4) Remove the tip from the liquid and transfer the tip into the container into which the liquid will be delivered. If any liquid remains on the tip, touch it to the side of the dispensing container or wipe gently with a lint-free tissue.

5) Touch the tip to the side of the receiving flask, and depress the plunger to the first stop. Wait 2 to 3 seconds, and then depress the plunger to the second stop to remove all of the liquid from the tip. Allow the plunger to return to its original position.

6) Remove tip from the end of the pipet.

Micropipetters are calibrated for aqueous solution but can be used with some organic solvents and solutions with densities different than water. Prior to use with an organic solvent, you should consult the manufacturer about compatibility of the tips and the pipetter with the solvent. With organic solvents, pipetting should be carried out rapidly, and the pipet should be opened after use to remove residual liquid or vapors.

Some solutions (organic solvents, serum, protein-containing solutions) can leave a residue in the tip. It is recommended that the tip be prewetted or rinsed with the solution prior to use. You can also use reverse pipetting. Reverse pipetting is the opposite of forward pipetting:

1) Select the appropriate micropipetter for your measurement (see Table 1.1 for reference). Using the volume adjustment knob, turn slowly until the indicator shows the desired volume. Always approach the desired setting from a higher volume to prevent mechanical errors that could affect accuracy. If you pass the desired setting, dial back up and then reset the volume.

2) Attach a new disposable pipet tip on the end of the pipetter. Make sure an airtight seal is created by pressing the tip gently onto the shaft. Avoid touching the narrow end of the tip.

3) Depress the plunger to the **second** positive stop. Holding the pipetter vertically or nearly vertical (within 20 degrees of vertical), immerse the tip a few millimeters into the liquid. Allow the plunger to slowly rise to its starting position. Liquid should rise in the tip. If you are drawing air into the tip, depress the plunger to remove the liquid/air and begin this step again, ensuring that the tip is immersed into the liquid.

4) Remove the tip from the liquid and transfer the tip into the container into which the liquid will be delivered. If any liquid remains on the tip, touch it to the side of the dispensing container or wipe gently with a lint-free tissue.

5) Touch the tip to the side of the receiving flask, and depress the plunger to the **first** stop. Do not blow out the remaining liquid. Some liquid should remain in the tip. Allow the plunger to return to its original position.

6) Return the remaining liquid to the original container (if appropriate). Remove tip from the end of the pipet.

In order to obtain precise and accurate measurements with pipetters, one must properly use and care for these instruments. Avoid drawing liquids or aerosols into the shaft. Pipette slowly and consistently, and always hold the pipette vertically. Never invert or lay a pipette down with liquid in the tip, even small droplets. Always use fresh tips to avoid contamination.

### *Burets*

Burets are long, narrow glass tubes with a stopcock on the end to control delivery of a liquid. A buret is calibrated and delivers precise volumes of liquid. In order for a buret to perform optimally, it must be properly cleaned. Prior to use, a buret should be rinsed several times with distilled water. After draining the final distilled water rinse, rinse the buret three times with the solution to be dispensed from the buret. After you are finished with the buret in your experiment, rinse it by filling it with distilled water and allowing it to drain.

Check the tip of the buret for air bubbles; the presence of air in the tip will result in inaccurate volume measurements. Air bubbles should be removed during the cleaning stage while distilled water is in the buret. Fill the buret about half full with water and open the stopcock. As the water drains through the tip, most air bubbles should be dislodged. If you are still having problems removing the air bubbles, gently shake the buret while the stopcock is open, or seek help from your instructor. Once the bubbles are removed from the tip, drain the liquid to within 5 cm of the bottom. Do not allow the buret to run dry after removing air bubbles.

Rinse the buret three times with about 10 mL of solution to be used and allow it to drain through the tip each time without allowing all of the solution to drain out. Then fill the buret with the solution to about 2-5 cm below the zero mark. Do not waste your time trying to get it exactly to zero. Once the buret is conditioned (cleaned, no bubbles in the tip) and properly filled, read and record the initial volume. Read at the bottom of the meniscus to avoid parallax errors, and record the volumes to two decimal places. Remember that the buret reading is the volume of liquid delivered from the glassware. When the buret is filled to the top line (0.00 mL), no liquid has been delivered. As the level of the liquid goes down, the volume measurement increases, and you now have a measure of the volume delivered. To find the volume delivered, subtract the initial volume (lower number) from the final volume (higher number).

## 1.10 ERRORS IN LIQUID MEASUREMENT

One of the most common errors in making liquid measurements is the failure to perceive the true bottom of the meniscus. It is often difficult to visualize the meniscus against a light background. You can improve your ability to read the meniscus by holding a card or paper behind the measuring device to darken the background.

A more serious error occurs when the meniscus is viewed at a level other than at eye level. The measurement error resulting from this is called the Parallax effect, the apparent displacement of an object depending on the position from which it is viewed. Figure 1.11 demonstrates the Parallax effect when reading a volume in a graduated cylinder. As demonstrated in the figure, the perceived volume differs depending upon the vantage point of the observer. For consistency and accurate measurements, always view liquid menisci at eye level as shown below.

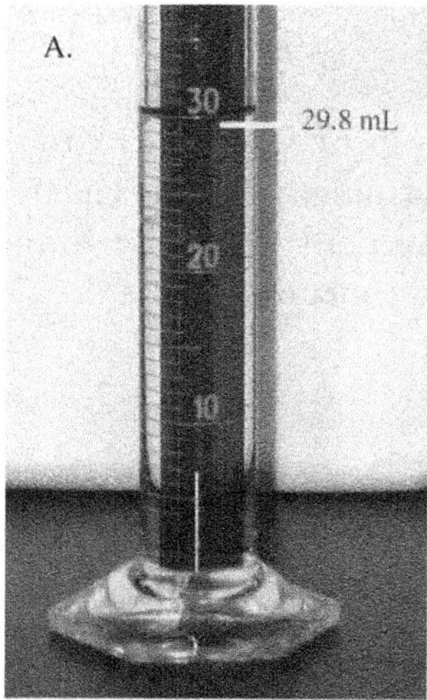


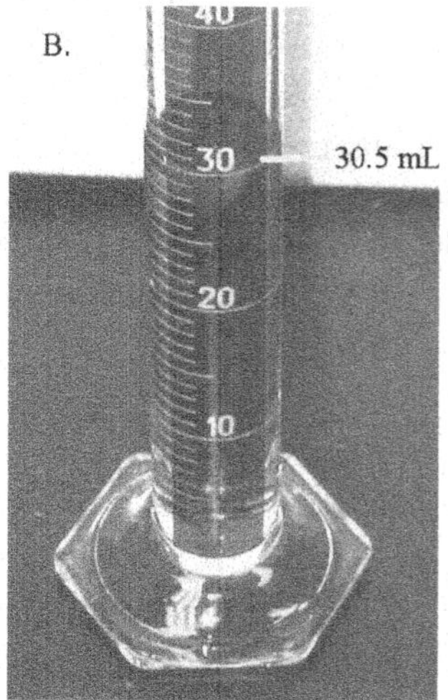


Figure 1.11 – Parallax effect when measuring liquids. The two images shown are the same graduated cylinder from two different angles: A) View of the graduated cylinder from close to eye level. The bottom of the meniscus is at 29.8 mL; B) View of the graduated cylinder from above. The bottom of the meniscus appears to be at 30.5 mL. The difference in the readings is due to the Parallax effect.

Finally, do not assume when preparing a solution of two liquids that the final volume of the solution will be equal to the addition of the volumes of the starting liquids. If you mix ethanol and water, the volumes are not additive. In fact, the volume of the solution will be less than that of the two separate volumes added together. This is due to the fact that a nonideal solution is created.

---

## Science in Action

### *Box 1.2 – Errors in Measurements*

It is important that scientists try to avoid (or at least minimize) errors in measurements, as these can lead to serious consequences. Scientists on two separate teams working on the Mars Climate Orbiter project failed to recognize an error in units (one team used English units whereas the other group used metric units). Due to an error in measurement (units), the spacecraft was lost by NASA in September 1999 (Recer, P. Associated Press, Nov. 10, 1999). Pharmacists also must be accurate in their measurements or patients may receive inappropriate doses of medication, which could have lethal consequences.

---

## 1.11 WEIGHING METHODS

The masses of substances can be obtained using balances. There are two types of balances, analytical and top-loading balances, that you will likely encounter in a laboratory setting (Figure 1.12). Analytical balances are accurate and precise instruments that measure masses to within 0.0001 g. They require locations free of drafts and vibrations that can cause errors in the measurements. A top-loading balance is used to measure substances with a precision of 0.001 g. For more accurate masses or smaller masses, the analytical balance is more precise and reduces errors.

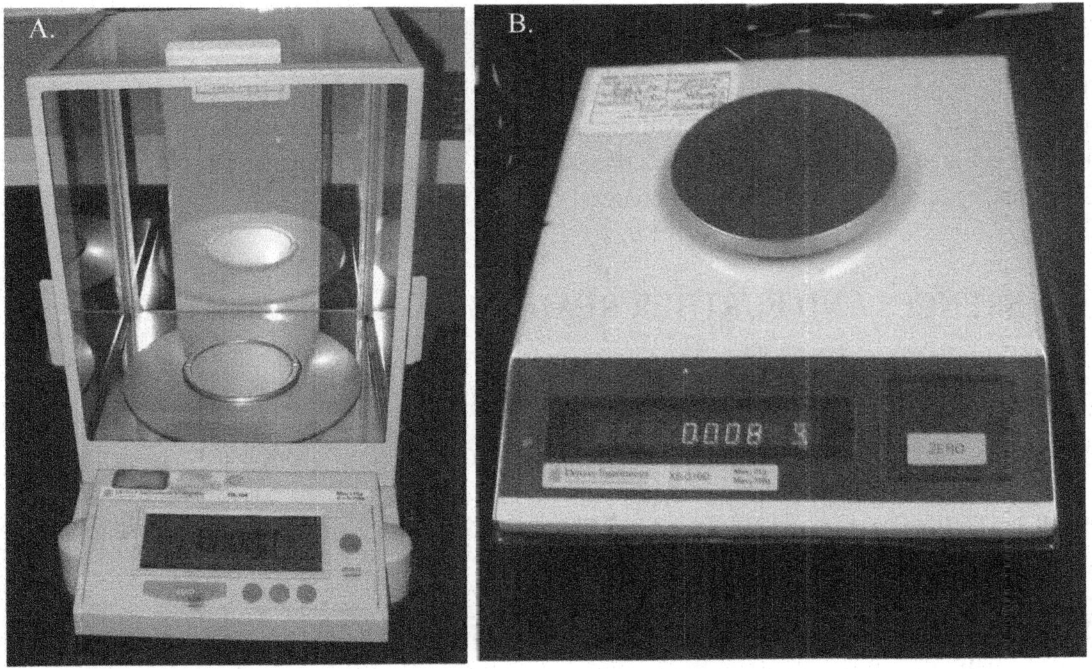

Figure 1.12 – Common Laboratory Balances: A) analytical balance, B) top-loading balance.

In order to ensure accurate measurements using an analytical balance, follow the procedure below:

1) Make sure the draft shields (doors) are closed and the balance is clean of debris or spills.

2) Turn on the balance and press the tare button until a stable zero reading occurs.

3) Open the doors and place the object inside, centering it on the pan as much as possible. Make sure the draft shields are closed before recording the object's mass. Allow the reading to stabilize before recording the object's mass.

Top-loading balances are operated similarly, except they do not have glass draft shields to open and close.

### Precautions when Weighing Substances

The following precautions should be followed to provide precise and accurate masses and to ensure a long life for the balance. You should always use weigh boats, weighing paper, or a container into which the sample is weighed to avoid damaging the surface of the balance pan. It is imperative that the mass of the weigh boat, paper, etc., is accounted for in order to receive an accurate mass of the desired substance. Do not weigh hot objects, convection heat will rise off of the object and inaccurate masses will be observed. Always allow an object to cool to room temperature before weighing to avoid this error. Samples that are extremely hygroscopic or volatile liquids should be weighed in a closed container or weighed rapidly. When you are finished, always clean the balance and area around the balance, close doors (if appropriate), and re-tare the balance.

Balances should be periodically calibrated to ensure precise measurements. The mass of standard weights should be checked to ensure that they are within the prescribed error range for the balance. If the masses are outside the control limits for precision, the balances should be serviced by a professional.

### Conclusion

Hopefully, after reading this chapter, you are prepared to have a safe and productive experience in the laboratory. Each future chapter will discuss background and additional safety concerns related to specific techniques or experiments. If you require additional information, please consult the suggested readings at the end of each chapter.

## REFERENCES AND FURTHER READING

### Scientific Writing

Dodd, J. S., Ed. *The ACS Style Guide*. Washington, DC: American Chemical Society, 1997.

Ebel, H. F., Bliefert, C., and W. E. Russey. *The Art of Scientific Writing*. Weinheim: Wiley-VCH, 2004.

Kanare, H. M. *Writing the Laboratory Notebook*. Washington, DC: American Chemical Society, 1985.

**Finding Chemical Information**

Bazler, J. A. *Chemistry Resources in the Electronic Age.* Westport: Greenwood Press, 2003.

Maizell, R. E. *How to Find Chemical Information.* 3rd ed. New York: Wiley-Interscience, 1998.

**Safety Information**

C. E. Gorman, Ed. *Working Safely with Chemicals in the Laboratory.* 2nd Ed. Schenectady, New York: Genium Publishing Corporation, 1995.

Hajian, H. G., and R. L. Pecsok. *Working Safely in the Chemistry Laboratory.* Washington, DC: American Chemical Society, 1994.

Safety in Academic Chemistry Laboratories. 5th ed. Washington, DC: American Chemical Society, 1990.

**Data/Statistical Analysis**

Grafen, A., and R. Hails. *Modern Statistics for the Life Sciences.* Oxford University Press: New York, 2002.

Skoog, D. A., West, D. M., and F. J. Holler. *Fundamentals of Analytical Chemistry.* 6th ed. Saunders College Publishing: Philadelphia, 1992.

**General Laboratory Techniques**

Coyne, G. S. *The Laboratory Handbook of Materials, Equipment, and Techniques.* Prentice-Hall: Englewood Cliffs, 1992.


**On-line References**

"Dr. Cal's Good Laboratory Practices." University of Illinois at Chicago. Jan 2005. <http://www.uic.edu/~magyar/Lab_Help/lghome.html>.

"Howard Hughes Medical Institute Laboratory Safety Program." Jan. 2005. <http://www.hhmi.org/research/labsafe

"Material Data Sheets on the Internet." Jan. 2005. <http://www.ilpi.com/msds/index.html >.

"Pocket Guide to Chemical Hazards and International Chemical Safety Cards." National Institute for Occupational Safety and Health (NIOSH). Jan. 2005. <http://www.cdc.gov/niosh/npg/npg.html>.

"Safety Links: Material Safety Data Sheets." Jan. 2005. <http://www.phys.ksu.edu/area/jrm/Safety/msds.html> .

# Experiment 1 – Use of Volumetric Glassware (Including calibrating micropipetters)

Biochemical experiments often utilize small quantities of materials; therefore, most measurements will be conducted using graduated pipettes and calibrated micropipetters. Before you start any laboratory work, it is a good idea to check the accuracy of the micropipetters. The purpose of this experiment is to learn how to accurately operate and check the operation of micropipetters.

## Determining Accuracy of the Glassware

You can assess the accuracy of the glassware by recording the weight of water delivered by the device. As water has known densities at specific temperatures, the volume of water actually delivered can be determined, and the accuracy of the glassware delivering it can be evaluated. If multiple measurements are recorded, the precision of the device can also be determined. For accurate determination of the density of water, record the temperature of the water used and look up the density of water at that specific temperature in a handbook.

## Prelaboratory Questions/Exercise

1) Design an experimental data table for your experiment. Make sure that your table includes appropriate columns such as volume on the micropipetter, mass of the delivered water, calculated volume of the water delivered.

2) Which pieces of delivering equipment would you predict are the most accurate and/or precise? Explain your predictions.

## Laboratory Experiment

1) Review the use of micropipetters and volumetric glassware in Chapter 1 of this text.

2) Conduct an experiment that determines the accuracy of the three different micropipetters (i.e. one calibrated for 0-20 µL, another for 0-200 µL, and one for 0-1000 µL) at measuring at least three different volumes of liquid (i.e., a pipette calibrated for 0-20 µL measuring 5 µL, 10 µL, and 20 µL).

3) Conduct an experiment that determines the precision of the three different micropipetters at measuring one of the volumes in Part 2 of the experiment.

4) Conduct an experiment that determines the accuracy and precision of a 100 mL beaker, a 100 mL Erlenmeyer flask, a buret, a 50 mL volumetric pipet, 50 mL volumetric flask and a disposable syringe at delivering 50 mL.

5) Record and report your findings in your notebook.

## Questions

1) Comment on the accuracy and precision of each micropipetter used in Parts 2 and 3.

2) Comment on the use of each piece of glassware in Part 4 to deliver 50 mL. Relate your findings to the information presented in Chapter 1. How do your results agree with your pre-laboratory predictions?

3) Is the error associated with each measurement systematic, random, or human? Explain your answer.

4) What could you do to improve the accuracy or precision of your measurements?

5) Would the results of the experiment change if you delivered a different liquid, such as ethanol?

# Acids, Bases, and Buffers

## 2.1 REVIEW OF ACIDS AND BASES

Acids and bases are substances that have opposite properties and reactivities. According to Arrhenius theory, acids are substances that, when added to water, produce hydronium ions, while bases produce hydroxide ions. Brönsted and Lowry defined an acid as a proton donor and a base as a proton acceptor. According to both these definitions, the dissociation of an acid (HA) in water to form its conjugate base ($A^-$) and hydronium ion is described by an acid dissociation constant ($K_a$):

$$HA + H_2O \rightleftharpoons A^- + H_3O^+$$

$$K_a = \frac{[H_3O^+][A^-]}{[HA]}$$

The larger the acid dissociation constant, the stronger the acid. In comparing acids, scientists often use $pK_a$ instead of $K_a$, where $pKa = -\log K_a$, because of the small dissociation constant of weak acids. The stronger the acid, the lower its $pK_a$. Similarly, a base dissociation constant ($K_b$) is used to describe the equilibrium reaction between a base (B) and water to form the conjugate acid (HB) and hydroxide ion ($B + H_2O \rightleftharpoons HB + OH^-$).

G.N. Lewis proposed an alternate definition for acids and bases. Lewis defined an acid as an electron pair acceptor and a base as an electron pair donor. Not only does this definition agree with the definitions previously proposed by Arrhenius and Brönsted and Lowry, but it also helps explain the acidic properties of a species such as $Al^{3+}$, which has no protons to donate.

Acids and bases can denature biological macromolecules, such as proteins and lipids. If your hands are exposed to an alkaline (strongly basic) solution, your skin may feel slippery. This is due to the base catalyzed hydrolysis of the lipids in the cellular membranes of the skin; therefore, care should be taken to avoid contact of the skin with strong bases or acids.

Science in Action

***Box 2.1 – Taste in Experimentation***
Historically, scientists would use all of their senses, including taste, in investigating substances in the laboratory. (However, you **do not** taste any substances in the laboratory). In the past, people would use taste to differentiate between acids and bases. For example, acids taste sour. The word "sauer" in German means acid. Sauerkraut is a sour cabbage that is created by soaking cabbage in lactic acid. In contrast, bases taste bitter rather than sour.

The strength of an acid or base can be measured by determining the amount of $H_3O^+$ or $OH^-$ that is formed in an aqueous solution. A convenient scale for describing acidity or basicity is the pH scale:

$$pH = -\log [H_3O^+]$$

A solution with a pH less than 7 is considered acidic. An alkaline solution has a pH greater than 7 and a neutral solution has a pH equal to 7 at room temperature.


## 2.2 MEASURING pH

The pH meter determines the pH of an aqueous solution by comparing the electrical potential (voltage) of the solution to a known reference potential. A pH electrode consists of two half-cells, one with an indicating electrode and another with a reference electrode. The potential of the working electrode is related to the concentration difference of $H_3O^+$ on either side of the glass surface of the electrode. The reference electrode produces a constant electrical potential. The electrical potential difference between the reference and working electrode is proportional to the concentration of the electrolyte measured, in this case $H_3O^+$. Many of the pH meters used today utilize a combination electrode that contains both half-cells within the same electrode. Some commercial glass electrodes consist of a glass tube with a thin-walled glass bulb on one end (see Figure 2.1). The tube contains a solution of fixed hydronium ion activity (concentration) and an internal reference electrode, generally a silver/silver chloride cell. When the glass bulb is immersed in a solution, an electrical potential develops across the glass surface in response to the hydronium ion activity (concentration) in the external solution. The potential difference is measured against the constant reference electrode potential, and the meter converts the potential to a pH value.

Figure 2.1 – An example of a glass combination pH electrode.

### *Common pH Meters*

Instruments that are used to measure pH come in various forms but have common operations. Figures 2.2 and 2.3 shows some common pH meters.

Figure 2.2 - Digital pH meters.

Figure 2.3 – An analog pH meter.

Although pH meters vary, most have similar features:

1) Temperature

The voltage produced by an electrochemical cell is dependent on the temperature of the solution. To control this variable, the temperature should be adjusted using the temperature knob. Typically, room temperature is 25°C.

2) Standardization/calibration

This function is used to calibrate the instrument to ensure accurate pH readings. The electrode is placed into a solution of known pH, and the pH reading is adjusted to the specified pH using this knob or button.

3) Function/mode

pH meters typically provide pH as well as millivolt measurements. (Recall that the pH meter measures a potential difference.) When the function or mode knob or button is adjusted to the pH setting, the instrument will display pH readings. In mV mode, the readings will be in millivolts, and the instrument can be used as a voltmeter.

## pH Meter Operations

### Electrodes

Care must be taken with pH electrodes. The pH electrode contains a thin surface or membrane of glass at the end of a tube. If handled improperly, the thin glass can break and render the electrode inoperable. The electrode should always be stored in saturated aqueous KCl solution or in a buffered solution. If the electrode dries out, it must be soaked in water for at least 24 hours prior to use.

Each time the electrode is used, it should be rinsed to prevent contamination by a solution. In order to rinse the electrode, carefully remove it from the storage container. Rinse the electrode with distilled or deionized water from a wash bottle into an empty beaker. Gently pat or blot the side of the electrode to remove any hanging water droplets with a Kimwipe®. Do not use excessive force or rub the electrode. When placing the electrode into a solution, use caution to avoid striking the bottom of the electrode against the sides or bottom of the container. Ensure that the tip or bulb of the electrode is completely immersed in the solution. Allow the electrode to equilibrate for 30-60 seconds prior to making a measurement.

*Standardization/Calibration of the Electrode*

Carefully lower the electrode into the standardized buffered solution of known pH. After equilibration, use a knob or button on the instrument to measure pH. If necessary, adjust the standardization/calibration knob until the pH display reads the pH of the standard solution. Some instruments have a calibration button that, when depressed, automatically adjusts the pH to the indicated pH. If the pH meter allows for calibration of multiple pHs, repeat the procedure on a rinsed electrode with a second or third solution. The instrument is now calibrated to ensure as much accuracy as possible.

## pH Paper

Another means to measure or approximate the pH of a solution is through the use of pH indicator paper. Litmus paper can be used to indicate simply if a solution is acidic or alkaline. Litmus is the oldest known pH indicator. It is naturally pink in color, but in acidic solutions it turns red, and in basic solutions litmus turns blue. Litmus paper is impregnated with the litmus in the acid or conjugate base forms. Acids will turn blue litmus paper red by donating a proton to the litmus and converting it to the protonated red form. Similarly, bases react with the litmus to convert it from the red form to blue, and, therefore, bases will turn red litmus paper blue. Other pH indicators exist, such as phenolphthalein or bromocresol blue, as well as commercially available pH indicator paper that allows for the approximate determination of pH of a solution. These papers are impregnated with one or a mixture of different pH indicators.

# 2.3 CONCENTRATION OF SOLUTIONS

In order to perform titrations and other biochemical manipulations, it is necessary to prepare solutions of appropriate concentrations and pH. There are numerous ways to describe the concentration of a solution, such as molarity (M), normality (N), and molality ($m$). Recall that a solution is composed of a solute (the substance in the smaller quantity) and the solvent (the substance in which the solute is dissolved).

The molarity of a solution is defined as the number of moles of solute per liter of solution. In most biochemical experiments, molarity is the preferred unit of concentration.

The normality of a solution is defined as the number of equivalents of solute per liter of solution. For acid-base reactions, an equivalent is the amount of substance that can accept or donate one mole of $H^+$. For a redox reaction, an equivalent is the amount of substance that can accept or donate one mole of electron. The normality can be related to molarity of a solution.

Acid-base reactions:

$$N = (\text{moles of } H^+ \text{ transferred}) \times M$$

Redox reactions:

$$N = (\text{number of electrons transferred}) \times M$$

The molality of the solution is a concentration unit that is not dependent on temperature, as volume is not included in the measurement. Molality is equal to moles of solute divided by kilograms of solvent.

Another commonly used concentration unit is mass (or weight) percentage of solute, which is mass of solute divided by mass of solution times 100.

The presence of ionized species in solution can affect properties such as solubility and boiling point. The most useful measure of total concentration of ions in solution is called the ionic strength ($\mu$).

$$\mu = 1/2 \sum_i c_i z_i^{\,2},$$

where $\qquad$ $c_i$ = concentration of each ion

$\qquad\qquad\qquad$ z = charge on an ion

For a 1:1 electrolyte, such as NaCl, containing cations and anions with charges of +1 and -1, respectively, the molarity equals the ionic strength. For 2:1 electrolytes, such as $Na_2CO_3$, the ionic strength is three times the molarity. For 3:1 electrolytes, like $Na_3PO_4$, the ionic strength is 6 times the molarity. Why does ionic strength matter? In solution, anions are surrounded by cations, and cations are surrounded by anions. This gives a partially positive charge in an atmosphere around the negatively charged anions and a partially negative atmosphere around the cations. The ionic atmospheres decrease the attraction between cations and anions and therefore affect properties such as solubility. The above-described situation ignores the impact of water and is just a theoretical interpretation of the impact of ionic atmospheres.

In order to account for the effects of ionic strength on the concentration of an ionic substance, the "effective concentration" or activity ($a$) is calculated as the product of the activity coefficient multiplied by the molar concentration. Therefore, the activity of substance x would be

$a_x = [x]\gamma_x,$

where $\qquad\qquad$ [x] = molar concentration of x

$\qquad\qquad\qquad$ $\gamma$ = activity coefficient of x

In an ideal solution, the species in solution would not interact with each other, and the concentration and the activity would be equal. In real solutions, the species interact to some degree, and this lowers the activity, which—in effect—lowers the observed concentration in solution. In dilute solutions ($\mu < 0.1$ M), ion concentrations are low; therefore, the ions are surrounded only by water molecules (and not other ions), and the solution behaves ideally. In an ideal solution, the activity coefficient is assumed to be 1. When ionic strength is larger than approximately 0.1 M, the ionic atmosphere affects the thermodynamic properties of the ion and corrections for the ionic strength must be made to the concentrations by correction factors called activity coefficients, which are a measure of how the system deviates from ideal.

Activity coefficients of ionic substances can be calculated using the Debye-Hückel equation:

$$\log \gamma = \frac{-0.51 z^2 \sqrt{\mu}}{1 + (\alpha \sqrt{\mu} / 305)},$$

where $\qquad\qquad$ $\gamma$ = activity coefficient of the ion

$\qquad\qquad\qquad$ z = charge on the ion

$\qquad\qquad\qquad$ $\alpha$ = effective diameter of the hydrated ion in nanometers

$\qquad\qquad\qquad$ $\mu$ = ionic strength of the solution

Table 2.1 provides values the effective diameter of the hydrated ion in aqueous solution at 25°C. These values can be used to calculate the activity coefficients of ionic substances.

Table 2.1 Effective Hydrated Diameters of Ions in Aqueous Solution (25°C)
(From J. Kielland, *J. Am. Chem. Soc.* **1937**, *59*, 1675).

| Cations | Anions | $\alpha$ (nm) |
|---|---|---|
| $Rb^+$, $Cs^+$, $NH_4^+$, $Tl^+$, $Ag^+$ | | 0.25 |
| $K^+$ | $Cl^-$, $Br^-$, $I^-$, $NO_3^-$, $HCOO^-$ | 0.3 |
| | $OH^-$, $F^-$, $HS^-$, $ClO_4^-$, $MnO_4^-$ | 0.35 |
| $Na^+$ | $HCO_3^-$, $H_2PO_4^-$, $HSO_3^-$, $CH_3COO^-$ | 0.4-0.45 |
| $Hg_2^{2+}$ | $HPO_4^{2-}$, $SO_4^{2-}$, $CrO_4^{2-}$ | 0.40 |
| $Pb^{2+}$ | $CO_3^{2-}$, $SO_3^{2-}$ | 0.45 |
| $Sr^{2+}$, $Ba^{2+}$, $Cd^{2+}$, $Hg^{2+}$ | $S^{2-}$ | 0.5 |
| $Li^+$, $Ca^{2+}$, $Cu^{2+}$, $Zn^{2+}$ $Sn^{2+}$, $Mn^{2+}$, $Fe^{2+}$, $Ni^{2+}$, $Co^{2+}$ | Phthalate$^{2-}$, $C_6H_5COO^-$ | 0.6 |
| $Mg^{2+}$, $Be^{2+}$ | | 0.8 |
| $H^+$, $Al^{3+}$, $Cr^{3+}$, $Fe^{3+}$, $La^{3+}$ | | 0.9 |
| $Th^{4+}$, $Zr^{4+}$, $Ce^{4+}$, $Sn^{4+}$ | | 1.1 |

## 2.4 TITRATIONS

When an acidic solution is mixed with one containing a base, a reaction occurs to form a salt and water. The overall acidity or basicity of the resulting salt solution will depend on the relative acidities and basicities of the starting materials. For instance, if one mole of a strong monoprotic acid reacts with one mole of a strong base, the resulting solution will be neutral. If equimolar amounts of a weak acid and a strong base react, the resulting solution will consist of the salt of the conjugate base of the weak acid, and this solution will have a basic pH. A reaction of a weak base with a strong acid will result in an acidic solution of the salt of the conjugate acid of the weak base.

Acid-base titrations can be used to determine the concentration of species in solution, the percentage of an acid or base in a mixture, or the $pK_a$ or $pK_b$ of a substance. Titration curves show the change in pH as an acidic or basic solution (titrant) is added to an initial solution. Figure 2.4 shows a titration curve generated by titrating a weak acid with a strong base. The equivalence point is clearly indicated, and the point in the buffer region where pH = pKa is also marked on the curve. The buffer region, where little change in the pH occurs, is found around the halfway point of the titration, typically at pH = $pK_a$ ± 1. (For further explanations of buffers, see Section 2.8.)

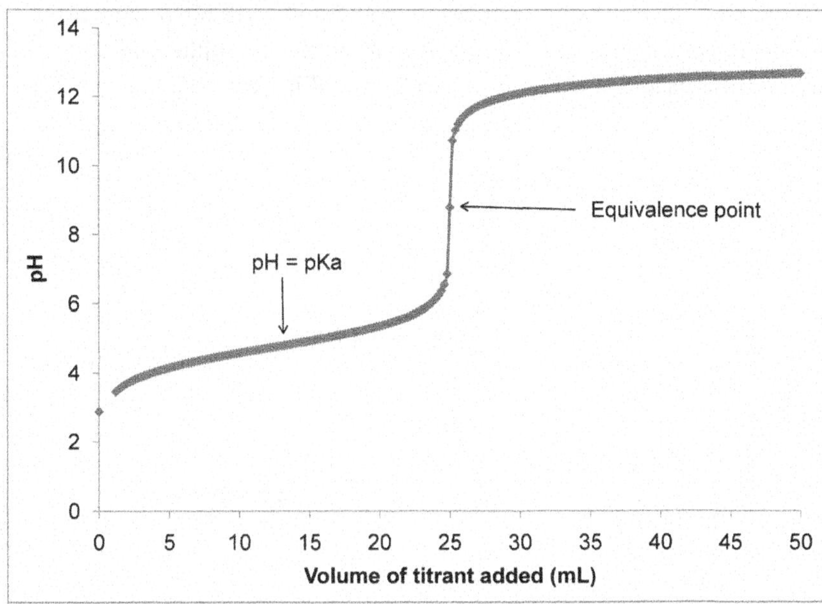


Figure 2.4 – Titration curve for a weak acid with a strong base.

The equivalence or endpoint of a titration (the point at which the moles of acid equals the moles of base) can be determined by evaluating the point in the curve (see Figure 2.4) in which the slope of $\Delta pH/\Delta V$ is greatest. This is often difficult to estimate by eye and the equivalence point can be determined easier by manipulation of the data. The first-derivative data (change in pH values divided by change in volume or $\Delta pH/\Delta V$) can be plotted versus volume of titrant. (See Figure 2.5 for an example.) The volume of titrant at which the maximum $\Delta pH/\Delta V$ occurs is the volume of titrant needed to reach the endpoint (in Figure 2.5, this would occur around 25.1 mL).

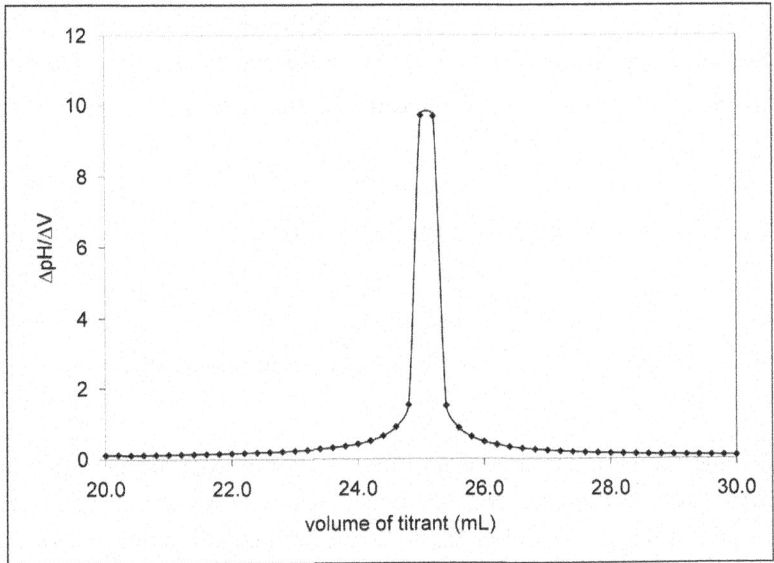


Figure 2.5 – First-derivative plot for the titration curve data in Figure 2.4.

Often, a better way to determine the endpoint of a titration is by using a Gran plot. Gran plots enable you to determine the endpoint with data taken before the endpoint rather than requiring data collected close to the equivalence point. As you can see in Figure 2.6, the endpoint can be extrapolated from the titration data. If the activity coefficients ($\gamma$) of the acid and its conjugate base in the solution are known, the $K_a$ can also be determined from the slope of the line on the Gran plot. In solutions of low ionic strength, the activity coefficient will approach unity, and often the equation is reduced to slope = $-K_a$. If the ionic strength varies during a titration, the Gran plot will not be linear. A discussion of activity coefficients and ionic strength can be found in Section 2.3.

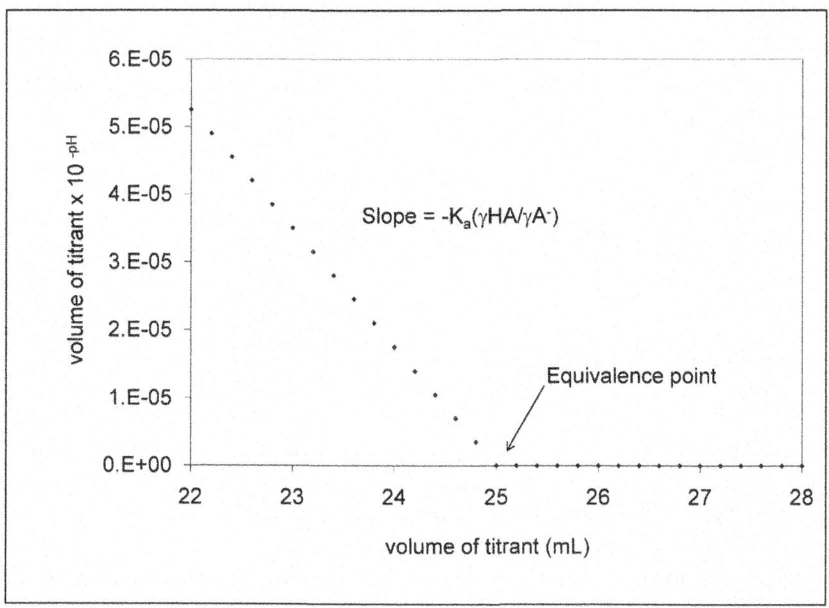


Figure 2.6 – Gran plot of the titration data from Figure 2.4.

## 2.5 AMINO ACIDS AND PROTEINS

Amino acids are the basic building blocks of proteins. All twenty naturally occurring amino acids have the same common functional groups attached to the alpha position of the carboxylic acid (COOH): a basic amino group ($NH_2$) and a hydrogen. Many also contain an ionizable side chain (where R contains a carboxylic group, amino group, hydroxyl etc.).

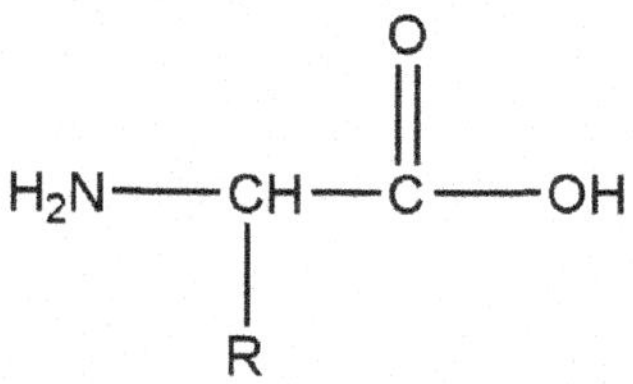

Figure 2.7 – Basic structure of an alpha-amino acid.

Knowledge of the acid-base properties of amino acids is crucial to understanding the structure and function of proteins. When reacted with acid or base, these amino acids change their ionic state, depending on the pH of the solution. For example, the amino acid with a nonionizable side chain pictured in Figure 2.8 can exist in three distinct ionization states. At a pH around 1, form I is present in which both the –COOH and the $NH_3^+$ group are in the acid form, resulting in a net charge of +1. Form II is a state referred to as a zwitterions, in which both the amino and carboxyl groups are ionized [$-NH_3^+$ and $-COO^-$]. The overall net charge on form II is zero. The pH at which the overall net charge on a molecule is zero is called the isoionic point. In electrophoresis (for more details, see Chapter 5), the isoionic point is referred to as the isoelectric point (pI), the pH at which a molecule will not migrate in an electric field. As the pH is raised above the isoionic point, the $NH_3^+$ group loses its proton, and form III is present. Therefore, these nonionized amino acids can be considered as diprotic acids, having two $pK_a$'s . One pKa represents the dissociation of the COOH, and the other represents the dissociation of the $NH_3^+$ group.

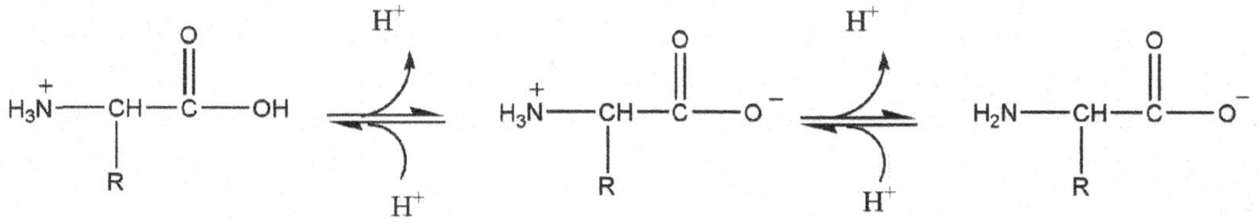

Figure 2.8 – Acid dissociation equilibria of an amino acid.

Some amino acids have a side chain that contains an additional ionizable functional group; these amino acids can be considered triprotic. Depending on the $pK_a$ of the side chain, its acid dissociation equilibrium will either occur directly after the ionization of the $\alpha$-COOH or after the $\alpha$-$NH_3^+$. Table 2.2 lists the $pK_a$ values for the twenty commonly occurring amino acids. When incorporated into proteins, the amino and carboxyl groups combine to form a peptide bond. Therefore, in proteins, the side chains of the amino acids have an especially important contribution to the charge/ionization state of the protein.

Table 2.2 – pKa Values for Common Amino Acids (adapted from Nelson, et al.).

| Amino Acid | $pK_a$ (COOH) | $pK_a$ ($NH_3^+$) | $pK_a$ (R group) |
|---|---|---|---|
| Alanine | 2.34 | 9.69 | |
| Arginine | 2.17 | 9.04 | 12.48 |
| Asparagine | 2.02 | 8.80 | |
| Aspartate | 1.88 | 9.60 | 3.65 |
| Cysteine | 1.96 | 10.28 | 8.18 |
| Glutamate | 2.19 | 9.67 | 4.25 |
| Glutamine | 2.17 | 9.13 | |
| Glycine | 2.34 | 9.60 | |
| Histidine | 1.82 | 9.17 | 6.00 |
| Isoleucine | 2.36 | 9.68 | |
| Leucine | 2.36 | 9.60 | |
| Lysine | 2.18 | 8.95 | 10.53 |
| Methionine | 2.28 | 9.21 | |
| Phenylalanine | 1.83 | 9.13 | |
| Proline | 1.99 | 10.96 | |
| Serine | 2.21 | 9.15 | |
| Threonine | 2.11 | 9.62 | |
| Tryptophan | 2.38 | 9.39 | |
| Tyrosine | 2.20 | 9.11 | 10.07 |
| Valine | 2.32 | 9.62 | |

## 2.6 TITRATION CURVES OF AMINO ACIDS

Titration curves can be used to determine the number of ionizable groups present in a sample, the $pK_a$ values of these groups, and ultimately the identity of an unknown acid, such as an amino acid. If an amino acid is dissolved in water and the pH is adjusted to between 1 and 1.5, all of the acidic functional groups will be fully protonated (form I). As a strong base, such as NaOH, is added to the solution through titration, the base will react with the acidic proton. After the reaction with the base, the relative amounts of form I and II (shown in Figure 2.9) will change. As the pH of the solution increases by the addition of NaOH, the amount of form II increases, as shown in Figure 2.9.

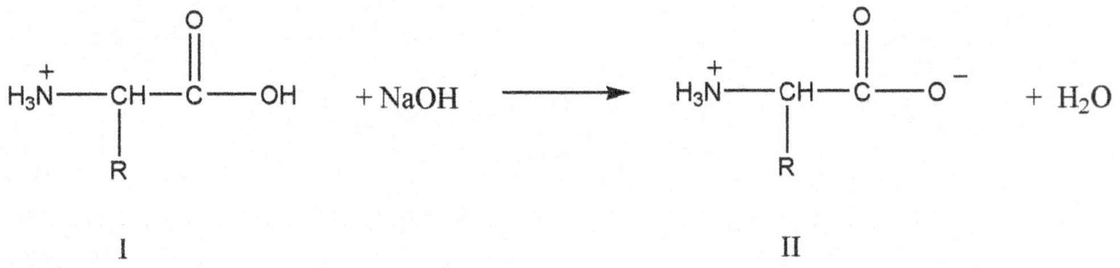

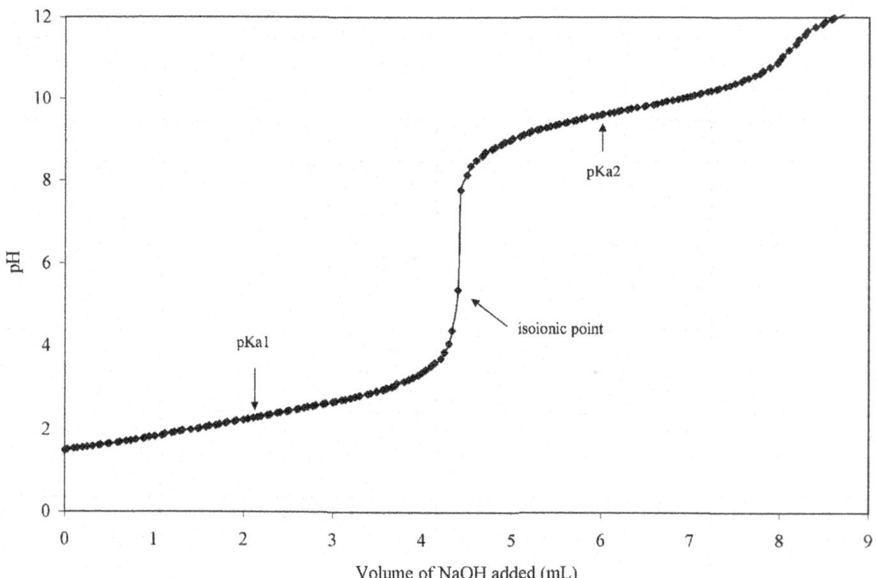

Figure 2.9 – Reaction of an amino acid with sodium hydroxide.

A titration curve can be created by plotting pH versus equivalents of NaOH added. An amino acid with a nonionizable side chain will have two ionizable groups; therefore the titration curve will contain one inflection point and two buffer regions. (See Figure 2.10.) The two buffer regions will occur around the pKa of the ionizable groups in the amino acid. (Refer to Table 2.2.)



Figure 2.10 –Titration curve of clycine with NaOH.

## 2.7 PREPARATIONS OF SOLUTIONS

Accurate preparation of solutions is important in laboratory experiments. It is important that the proper quantities of substances be dissolved in the correct amount of solvent. If the solute is a pure liquid or solid, a solution is created by weighing or measuring the correct mass or volume of the solute and dissolving the solute in the solvent. The volume is adjusted with solvent until the desired volume of solution is reached. To prepare a 0.030 molar solution of glucose, one would weigh out 5.4 g of glucose and dissolve it in enough water to make 1.0 L of solution. If you add 1.0 L of water to the 5.4 g of glucose, you will not create a 0.030 M solution because the total volume of the solution will be greater than 1.0 L.

Frequently, one will have to create solutions by diluting a more-concentrated solution. A useful equation for dilution is

$$M_1 V_1 = M_2 V_2,$$

where $\quad$ $M_1$ = the concentration of the original (more-concentrated) solution

$V_1$ = the volume of original solution

$M_2$ = the concentration of the dilute solution

$V_2$ = the final volume of the solution after dilution

For example, 500 milliliters of a 0.2 M solution of NaCl can be created from a 1.0 M NaCl solution, by diluting 100 mL of the 1.0 M solution with enough water to create 500 mL of solution.

It is also important to remember that masses and moles are additive, but concentrations are not. For example, if a container contains 0.5 g glucose and another 1.0 g of glucose was added, one could deduce that the flask contains 1.5 g of glucose. However, if a 0.2 M solution is mixed with 1.0 M solution, the resulting solution is not 1.2 M. Because of dilutions, the molarity of the resulting solution will be somewhere between 0.2 M and 1.0 M, depending on the volumes of solutions used.

Most of the solutions prepared in the biochemistry laboratory contain more than one solute, e.g., 1 liter of 0.1 M KCl, 50 mM imidazole at pH 7.2. You might think that dissolving 7.45 g of KCl in a 1 L volumetric flask and dissolving 3.45 g of imidazole in a second 1 L flask, combining the two solutions, and adjusting the pH to 7.2 would produce the required solution, but this is incorrect. This procedure creates a solution that is approximately 0.05 M KCl and 25 mM imidazole.

The correct way to prepare this solution is to dissolve 7.45 g of KCl and 3.45 g of imidazole in the same flask with approximately 750 mL of water. The pH is adjusted to 7.2 with either acid (usually 6 M HCl) or base (usually 6 M NaOH). The volume is then increased to 1 L with water.

## 2.8 INTRODUCTION TO BUFFERS

Many enzymatically catalyzed biochemical reactions occur in only a narrow pH range or are sensitive to changes in pH. Deviations in pH can also alter the ionization state of the side chains of amino acids in proteins. Changes in the ionization state can alter enzymatic activity, especially if the side chain is involved in covalent catalysis or acts as an acid or a base. This altered ionization state can also affect the folding and shape of the protein. Therefore, controlling the pH of the solution is very important. Buffers enable careful control of pH in biochemical experiments.

Buffers are solutions that resist large changes in pH. Buffers contain a weak acid (or base) and its conjugate in near equal concentrations. Buffers are most effective when the pH of the target solution is within one pH unit of the buffer's $pK_a$. It is therefore important that the appropriate buffer be carefully selected for each biochemical experiment. For example, if the pH decreases during a biochemical reaction, a buffer with a $pK_a$ slightly lower than the desired pH should be chosen. Conversely, if the pH increases during the experiment, a buffer with a $pK_a$ slightly greater than the desired pH is selected. Table 2.3 shows common buffers used in biochemical experiments. In addition, amino acids (see Table 2.2) can be used as buffers.

Table 2.3 – Commonly Used Buffers for Biochemical Experiments.

| Name | $pK_a$ | | Name | $pK_a$ |
|---|---|---|---|---|
| Phosphate $H_3PO_4$ / $NaH_2PO_4$ $NaH_2PO_4$ / $Na_2HPO_4$ $Na_2HPO_4$ / $Na_3PO_4$ | 2.15 7.20 12.15 | | Imidazole $C_3N_2H_5^+$ | 7.05 |
| Glycine $^+H_3NCH_2COOH$ | 2.34 9.78 | | N-2-hydroxyethyl-piperazine-N'-2-ethanesulfonic acid [HEPES] | 7.55 |
| Citrate $HOOCCH_2C(COOH)$ $OHCH_2COOH$ | 3.09 4.75 5.41 | | Tris(hydroxymethyl) aminomethane [Tris] | 8.10 |
| Formate $HCOOH$ | 3.77 | | N-tris(hydroxymethyl) methylglycine [Tricine] | 8.15 |
| Succinate $HOOCCH_2CH_2COOH$ | 4.19 5.48 | | N,N-bis(2-hydroxy-ethyl) glycine [Bicine] | 8.35 |
| Acetate $CH_3COOH$ | 4.76 | | Borate $H_3BO_3$ | 9.23 |
| Pyridine $C_5NH_6^+$ | 5.14 | | Bicarbonate $HCO_3^-$ | 10.3 |
| 2-(N-Morpholino) ethanesulfonic acid [MES] | 6.15 | | Triethylamine [TEA] | 10.65 |
| Piperazine-N,N'-bis (2-ethanesulfonic) acid [PIPES] | 6.8 | | | |

The Henderson-Hasselbalch equation can be used to demonstrate the mathematical relationship between pH of a solution and the buffer's $pK_a$.

$$pH = pK_a + \log([A^-]/[HA]),$$

where     pH = the pH of the solution

          $pK_a$ = the $pK_a$ of the weak acid in the buffer

          $[A^-]$ = the concentration of the conjugate base form of the acid in solution

          [HA] = the concentration of the undissociated acid remaining in the solution

Since the $pK_a$ of the buffer is virtually constant, the target pH depends on the ratio of the concentration of $A^-$ to the concentration of HA rather than on the total concentration of the buffer. If the concentrations of $A^-$ and HA are equal, the pH of the solution equals the $pK_a$ of the acid. (See buffer regions of titration curves in Figures 2.4 and 2.10.) In a solution of equal concentration of a weak acid-base conjugate pair, the buffering capacity, i.e., the resistance to changing pH, is greatest.

Buffer capacity is the ability of a buffer to resist changes in pH. It can be defined in two ways: 1) the number of moles of acid ($H_3O^+$) or base ($OH^-$) that can be added to the 1.0 L of a buffer to change the pH by 1 unit or, 2) the pH change that occurs when a specific amount of acid or base is added to a buffer. The former definition is more frequently used because it can be applied to any buffer at any concentration.

## 2.9 PREPARATION OF BUFFERS

There are many ways to prepare a buffer. All of these methods involve adjusting the ratio of acid [HA] and conjugate base [$A^-$] forms (or base and conjugate acid) of the compounds in the solution to give the desired pH. Unless otherwise indicated in a procedure, buffers should be prepared with deionized or ultrapure water.

1) **The one component method.** Dissolve the acidic form of the compound in about 60% of the water required for the final solution. Adjust the pH with a strong base, such as NaOH (or one that contains the same ions as the buffer components). Add enough water to reach 95% of the final volume of the solution. Check the pH and adjust with NaOH or HCl, if necessary. Add water to reach the final volume. This creates a buffer composed of the acid HA and its conjugate A- by acid equilibrium.

   For a buffer composed of a base and its conjugate, dissolve the base in about 60% of the water required to make the solution. Adjust the pH with a strong acid, such as HCl (or one that contains the same ions as the buffer components). Add enough water to adjust the volume to 95% of the final volume of the solution. Check the pH and adjust, if necessary. Adjust the volume to the final volume.

   This method may require large amounts of acid or base (HCl or NaOH) and may take a long time to reach the final pH. It is also easy to overshoot the pH or the volume. With this method, the ionic strength of the final solution is unknown.

2) **The mathematical method.** Using the buffer $pK_a$ (or $pK_b$) value for the compound and the Henderson-Hasselbalch equation, calculate the mass of acid and its conjugate (or base and its conjugate) present in the buffer at the desired pH. Weigh out the amount of acid (or base) and the salt of the conjugate that you calculated. Dissolve both in the same flask in the appropriate amount of water to create slightly less than the desired volume of the final solution. Check the pH and adjust by adding small quantities of NaOH or HCl, if necessary. Add water to achieve the final volume. Using this method, one rarely needs to adjust the pH, and the ionic strength can be easily calculated.

3) **The two solution method.** Prepare separate solutions of the acid (or base) and salt form of the conjugate. The solutions should have the same concentration of buffer (and ionic

strength – if this is important) as is desired in the final solution. Mix one solution with the other until the desired pH is obtained. In this method, the ionic strength can also be calculated. An advantage is that you can create buffers of identical ionic strength with varying pHs. One disadvantage is that you may waste solutions because you may not use all of the solutions prepared to create the buffer.

## REFERENCES AND ADDITIONAL READINGS

Clark, J. M., and Switzer, R. L. *Experimental Biochemistry*. 2nd ed. New York: W.H. Freeman and Company, 1977.

Coyne, G. S. *The Laboratory Handbook of Materials, Equipment, and Techniques.* Englewood Cliffs: Prentice-Hall, 1992.

Farrell, S. O., and R. T. Ranallo. *Experiments in Biochemistry*. United States: Thompson Learning, 2000.

Nelson, D. L., and M. M. Cox. *Lehninger Principles of Biochemistry*. New York: Worth Publishers, 2000.

Sambrook, J., and D. W. Russell. *Molecular Cloning. A Laboratory Manual.* 3rd ed. New York: Cold Spring Harbor Laboratory Press, 2001. Appendix A.

Skoog, D. A., West, D. M., and F. J. Holler. *Fundamentals of Analytical Chemistry*. 6th ed. Philadelphia: Saunders College Publishing, 1992.

# Experiment 2. Preparing Solutions, Measuring pH, and Working with Buffers

## Purpose of Experiment

In this experiment, you will prepare a phosphate buffer at a pH of 8.0 using two different methods (described in Chapter 2) and determine the easiest and most effective way to prepare the buffer. In part two of the experiment, you will determine the buffer capacity of your buffers by titrating them with strong acid and base and comparing the buffer capacity to that of an unbuffered solution. Alternatively, students could prepare the phosphate buffer to be used in Experiment 6 (Spectrophotometric Studies of Nucleic Acids in Solution).

## Prelaboratory Assignment

Prior to arriving in lab, each student should plan the experiments to be conducted. The procedures below provide guidelines for the preparation of samples, but the student should calculate the volume or amounts of each substance to use for each experiment. These amounts should be determined and written in the laboratory notebook before the start of the experiment. If required by your instructor, have these quantities verified before beginning experimentation.

## Materials

$Na_2HPO_4$ (dibasic sodium phosphate)
$NaH_2PO_4$ (monobasic sodium phosphate)
NaCl (0.50 M solution)
HCl (0.50 M solution and 6 M solution)
NaOH (0.50 M solution and 6 M solution)
Buret
Buret stand
Balance
pH meter
Spatula
Beakers and Erlenmeyer flasks
Volumetric flasks (250 mL)
Wash bottle

## Safety

Wear goggles. Use caution when working with strong acids and bases. Dispose of the solutions at the end of the experiment in the proper manner, as indicated by the instructor.

## Experimental Methods

1) Prepare 250 mL of 0.020 M phosphate buffer (pH 8.0) using <u>two</u> of the methods described in Chapter 2. One of your methods should include the two solution method. For that method, you should prepare 50 mL of 0.20 M $Na_2HPO_4$ (dibasic) and 25 mL of 0.20 M $NaH_2PO_4$ (monobasic). Slowly add the $NaH_2PO_4$ to the $Na_2HPO_4$ until the pH is 8.0. Dilute the 0.20 M buffer to 0.020 M for a final volume of 250 mL.

2) Rinse and fill one buret with 0.50 M HCl and a second buret with 0.50 M NaOH.

3) Familiarize yourself with the operations of the pH meter, and calibrate the pH meter.

4) Transfer 50 mL of one of your buffered solutions to a beaker. Place a stir bar in the bottom of the beaker. Position the beaker on a stir plate (located near the pH meter). Carefully place the pH electrode in the solution, so that it is covered with solution but cannot be damaged by the stir bar. Record the pH of the solution.

5) Record the initial buret reading. Slowly add HCl (e.g., 0.05 – 0.1 mL increments), and record the buret reading and the pH of the solution after each addition. As the change in pH slows, you may want to increase the increments of HCl added. Continue adding HCl until the pH reaches approximately 2.0.

6) Repeat Steps 4-5, but titrate the solution with NaOH instead of the HCl. Add the NaOH until the pH reaches approximately 12.0.

7) Repeat Steps 4-6, but use 50 mL of 0.50 M NaCl instead of the buffered solution.

## Questions

1) Compare the two methods of preparing the buffers. Include a discussion of which method of preparing the buffers was easier, faster, more economical.

2) Over what pH range is your buffer effective? (What is the buffering capacity for your solutions?)

3) Can the NaCl solution serve as a buffer? Explain your answer.

4) Determine the buffer capacity for your buffer.

5) Determine the $pK_a$ values for the buffer components. Compare them with the literature values given in the introduction (Table 2.2).

# Experiment 3. Preparation of a Multiple-Component Solution

## Purpose of Experiment

In this experiment, you will prepare the Bradford reagent for Experiment 5. This solution contains multiple components and must be prepared as described. The accuracy and precision of the solution preparation can be verified spectrophotometrically. The verification can be conducted by the students (see Chapter 3 for more details) or by the instructor or an instructor's aid.

## Materials

Coomassie Brilliant Blue G-250
95% ethanol
85% (w/v) phosphoric acid
Whatman #1 filter paper

## Safety

Wear goggles. Use caution when working with strong acids and bases. Dispose of or store the solutions at the end of the experiment in the proper manner, as indicated by the instructor.

## Pre-Laboratory Assignment

Complete the following exercise before lab and submit it to your instructor. Describe in detail how you would prepare 200 mL of the following solutions:

1) BCA Reagent A (1% BCA, 2% sodium carbonate, 0.16% sodium tartrate, 0.4% sodium hydroxide, and 0.95% sodium bicarbonate; adjust pH of solution to 11.25 with NaOH or $NaHCO_3$).

2) Lowry Reagent 1: Mix one volume of Reagent B (0.5% copper sulfate pentahydrate, 1% sodium or potassium tartrate) with 50 volumes of Reagent A (2% sodium carbonate, 0.4% NaOH). Reagent A and B can be stored indefinitely at room temperature. NOTE – you will need to describe how reagents A and B are created individually and then describe how Lowry Reagent 1 is created.

3) Enzyme buffer (0.1 M Tris-HCl (pH 9.5), 0.1 M NaCl, 5 mM $MgCl_2$).

## Laboratory Assignment

1) Prepare 100 mL of Bradford working buffer by dissolving 10 mg Coomassie Brilliant Blue G-250 in 5 ml 95% ethanol. Add 10 ml 85% (w/v) phosphoric acid. When the dye

has completely dissolved, dilute to 100 mL. Filter the solution through Whatman #1 or #2 paper.

2) To verify solution preparation, create a mixture of 50 μL of water and 2.5 mL of Bradford reagent.

3) Turn on the spectrophotometer and follow instructions to use the spectrophotometer.

4) Blank the instrument with water in quartz cuvettes between 400 and 700 nm.

5) Scan the sample from Step 2 in the quartz cuvette between 400 and 700 nm. Rinse the cuvettes with alcohol (methanol or ethanol) and then water to remove any residual dye.

## Questions

1) Describe the appearance and preparation of your solution.

2) How does the spectrum of your reagent compare with the one in Figure 3.15 (Chapter 3 material)? If differences exist, offer possible experimental explanations for the variation.

3) If you were to prepare the solution again, what differences would you make in your experimental procedure?

# Experiment 4. Ionic Properties of Amino Acids and Peptides

## Purpose

You will be provided with three unknown samples (a dipeptide and the free amino acids found in that dipeptide). The amino acids may be in any functional state (i.e., zwitterionic form, free base form, amino acid hydrochloride). The purpose of this experiment is to titrate your sample with HCl and NaOH to determine the molecular weight of the samples, the $pK_a$ values for all ionizable functional groups in the samples, and the identity of the samples.

## Prelaboratory Questions

1) Consider the titration of 100.0 mL of 0.0200 M glutamic acid with 0.121 M NaOH. Calculate the volume of base that must be added to reach each of the equivalence points.

2) Sketch with approximate pH values and equivalence of base, the titration curve for the dipeptide Gly-Asp.

## Materials

Unknown samples
Standardized HCl (between 1 and 2 M)
Standardized NaOH (between 1 and 2 M)
pH meter
Stir plate and stir bar
Buret
Buret stand
Wash bottle
Beakers
Standard buffers
Balance
Spatula
Weigh paper or boats
Deionized water

## Safety

Wear goggles. Use caution when working with strong acids and bases. Dispose of the solutions at the end of the experiment in the proper manner, as indicated by the instructor.

## Experimental Methods

1) Rinse and fill one buret with standardized HCl and a second buret with standardized NaOH. Record the concentration of the acid and base.

2) Familiarize yourself with the operations of the pH meter, and calibrate the pH meter.

3) Weigh about 100 mg of your unknown sample, recording its mass to the nearest milligram.

4) Dissolve the sample in 20 mL of deionized water in a 50 mL beaker. Place a stir bar in the bottom of the beaker. Place the beaker on a stirring plate (located near the pH meter). Carefully position the pH electrode in the solution so that it is covered with solution but cannot be damaged by the stir bar. Record the pH of the solution. NOTE – instead of using a stir bar, the solution can be carefully shaken after each addition.

5) Record the initial buret reading of HCl. Slowly add HCl until the pH of the solution reaches a pH of 1.5 and until the entire unknown sample is dissolved. Record the final buret reading for the HCl.

6) Record the initial buret reading of NaOH. Slowly add NaOH (in 0.05 – 0.1 mL increments) to the solution from Step 5, and record the buret reading and the pH of the solution after each addition. Continue adding NaOH until the solution reaches a pH of 12.0.

7) Repeat Steps 3-6 with the other samples as assigned by your instructor.

## Alternative Additional Exploration

Repeat the above-described experiment with a tripeptide or tetrapeptide and the free amino acids that compose the peptide.

## Questions

1) Determine the $pK_a$ values for each ionizable group in your samples.

2) Using the weight of your solid sample and the titration data, determine the molecular weights of your samples. You will want to use graphical manipulation of the data (such as first derivative plot or Gran plot) to more accurately determine equivalence points and pKa.

3) What is the identity of each of your unknown samples?

4) How do your results compare with literature data for your samples?

5) How do the pKa values change for the ionizable side chains when the amino acids are bound into a dipeptide? Do your results agree with literature results?

# Spectrometric Methods

## 3.1 INTRODUCTION TO SPECTROSCOPY

Spectroscopy is the study and interpretation of the interaction of radiation (such as electromagnetic radiation) or energy with matter. Spectrometry is the measurement of radiation with a photoelectric transducer or other electrical device. There are a wide range of spectrometric techniques, including atomic absorption and atomic fluorescence spectrometry, mass spectrometry, ultraviolet/visible molecular spectrometry, Raman spectroscopy, infrared spectrometry, and nuclear magnetic resonance spectroscopy. This chapter will cover the basics of atomic and molecular spectroscopy and will focus on the commonly used techniques in biochemical laboratories.

Electromagnetic radiation (light) is generally viewed as a propagating electric and magnetic field; in this sense, electromagnetic radiation is seen as wave-like. Many properties of this radiation, however, rely on it having discrete packets of energy called photons. Therefore, light has both a wave and a particle nature.

According to Planck's theory, the energy of a photon (E) is related to the frequency of the wave ($\upsilon$):

$$E = h\upsilon,$$

where h = Planck's constant = $6.626 \times 10^{-34}$ J·s

Frequency can also be related to wavelength ($\lambda$) through the speed of light (c = $3.00 \times 10^8$ m/s):

$$c = \lambda\upsilon$$

Short wavelengths correspond to high-energy radiation, while longer wavelengths are associated with lower energies. Figure 3.1 shows the regions of the electromagnetic spectrum.

Spectroscopic methods detect specific interactions between electromagnetic radiation of distinct energies and matter. These interactions depend on the energy of the radiation and have different effects on molecules, including electron excitation, molecular vibrations, and nuclear spin flips. Figure 3.1 lists the regions of the electromagnetic spectrum that are important for each analytical spectroscopic method and indicates the quantum transitions associated with those regions of the spectrum.

Science in Action

***Box 3.1 – Why are trees green?***

Colors arise in biological tissues and substances because they contain chromophoric materials or pigments. Photosynthetic organisms, such as plants and trees, contain numerous pigments, including chlorophyll. Chlorophyll appears green to the human eye because it absorbs red and blue light and emits green light, which is detected. Other naturally occurring pigments have different colors: ß-carotene in carrots is orange; lycopene in tomatoes is red.

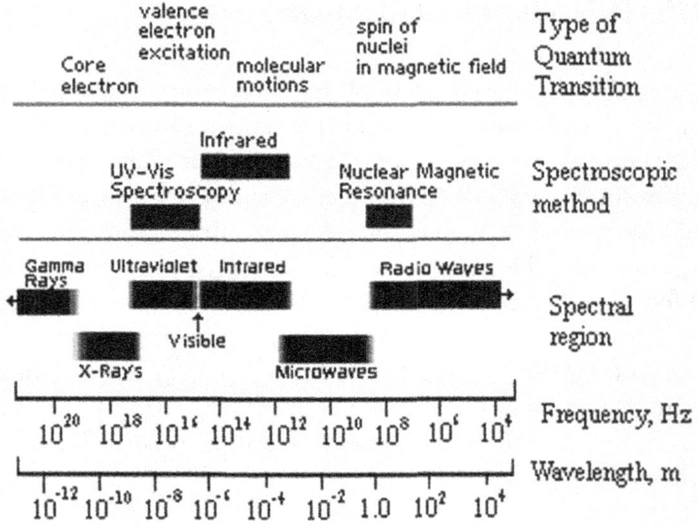


Figure 3.1 – Regions of the electromagnetic spectrum.

# 3.2 ULTRAVIOLET-VISIBLE ABSORPTION SPECTROMETRY

Ultraviolet-Visible (UV-Vis) Absorption Spectrometry is an analytical technique that measures the wavelength-dependent absorption of light in the ultraviolet and visible regions of the electromagnetic spectrum. It is useful for the identification and quantification of various organic and inorganic substances.

Absorption spectroscopy relies on two power measurements: the radiant power in a light beam after it has passed through a sample (P) and the power of the incident beam ($P_0$). Radiant power, or intensity, is a measure of energy of the electromagnetic radiation reaching a detector per unit of time, usually measured in watts. Absorption spectroscopic techniques involve two terms based on the ratio of P to $P_0$, namely transmittance and absorbance.

Transmittance (T) of a sample is the ratio of power of the transmitted beam to the power of the incident beam:

$$T = P/P_0$$

For most experimental analyses, transmittance becomes the ratio of radiant power transmitted through a sample (analyte) versus radiant power transmitted through a blank (usually the solvent):

$$T = P_{sample}/P_{blank}$$

Transmittance is also often expressed as a percentage instead of a ratio.

Absorbance is defined as

$$A = \log(1/T) = -\log T = -\log(P/P_0) = -\log(P_{sample}/P_{blank}) = -\log T$$

For monochromatic radiation that passes through a sample of constant concentration, the absorption is directly related to the optical path length and the molar concentration. This relationship is known as the Beer-Lambert law (or Beer's law):

$$A = abc,$$

where   A = absorbance
a = proportionality constant known as absorptivity (L/g·cm)
b = optical path length in centimeters
c = concentration in g/L

When the concentration is expressed in moles/L, the proportionality constant ($\varepsilon$) is called the molar absorptivity, or molar extinction coefficient, and has units of (L/mol·cm):

$$A = \varepsilon bc$$

Beer's law reveals a linear relationship between concentration of a solution and absorbance. This relationship allows absorption spectroscopy to be used for determining the quantitative concentration of an analyte in solution or for monitoring changes in concentration. There are, however, some deviations from the linearity of Beer's law that occur even when b is constant. Such deviations result from high analyte concentrations and molecular interactions, chemical changes to the structure of the compound during analysis, and the use of polychromatic light.

### Standard Curves

If a system obeys Beer's law, spectroscopy can be used quantitatively with a known molar extinction coefficient and measured absorbance to determine the concentration. If the molar extinction coefficient is not known and a sample of the known concentration is available, Beer's law can still be used by constructing a standard curve. A series of samples of known concentrations or amounts (e.g., mg/mL) are prepared, and the absorbance is obtained. Figure 3.2 shows an example of a standard curve. Once the data are collected, a plot of absorbance versus concentration (or amount) is created and used to determine the concentration of an unknown whose absorbance is known. In order to be accurate, the absorbance of the unknown must fall within the linear limits of the standard curve.

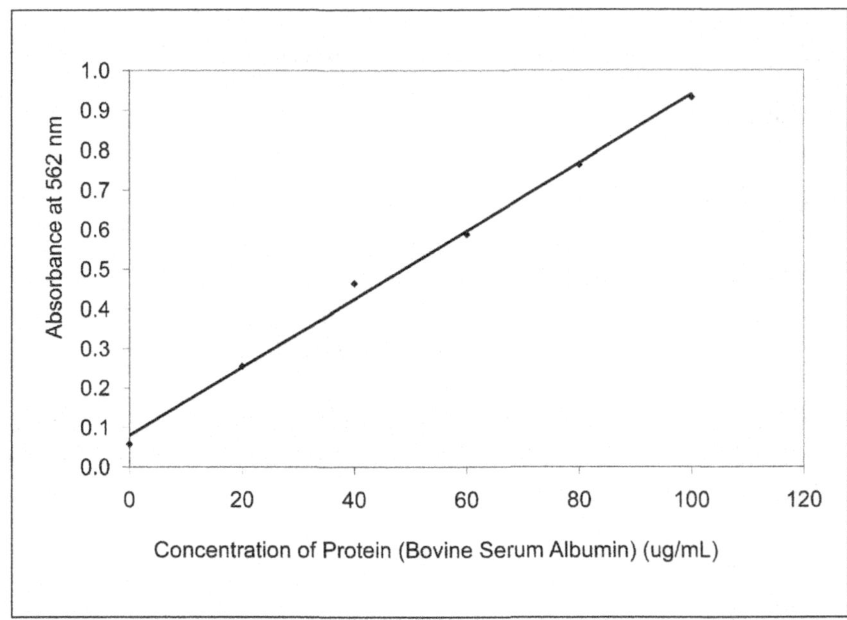


Figure 3.2 – Example of a standard curve for a bicinchonic acid protein assay.
(See Section 3.7 for more details about the assay.)

## Spectrometers

As indicated above, most absorbance (or transmittance) measurements involve the ratio of the light absorbed (transmitted) by the sample versus the light absorbed (transmitted) by a blank. Commercially available spectrophotometers come in various designs. Single-beam instruments require that the reference sample, typically a cuvette (sample holder) containing the solvent, be scanned first. The spectrum of this blank is stored by the computer interfaced with the instrument. The sample of interest is then placed in the sample chamber and scanned. The absorbance of the sample solution is calculated using the stored reference data ($Abs = -\log(P/P_0)$).

Other spectrophotometers are double-beam instruments. The monochromatic light is split by a beam splitter so that it passes simultaneously through a sample and a reference. The instrumentation determines the ratio of radiant power absorbed by the reference versus the sample through a series of optics and a detector—frequently a photomultiplier tube or photodiode.

## Basic Use of Spectrophotometers

Although, each individual spectrophotometer is different, some basic operating procedures are followed when using any instrument:

1) Turn on the instrument and allow it to warm up according to the manufacturer's specified time.

2) Set the instrument to a specific wavelength or wavelength range over which the absorbance will be recorded.

3) Ensure that the cuvette(s) are clean and unscratched. For a dual-beam instrument, the cuvettes should be used in a matched set. Do not mix and match cuvettes. Whenever handling cuvettes, avoid touching surfaces where light will pass through, and always wipe the cuvettes with a Kimwipe®, lens paper, or soft cloth to clean without scratching the surface.

4) Fill the cuvette with enough solution so that the solution height is higher than the level where light will pass through the cuvette. The initial solution in the cuvette should be the blank, which is the solvent or solution containing everything except the analyte of interest. Wipe the surface with a Kimwipe® or lens paper.

5) Blank the instrument by measuring the absorbance of the cuvette with the blank solution. Some instruments have a blank button or zeroing program.

6) Remove the cuvette and replace the blank with the solution to be measured. Wipe the outside of the cuvette with a Kimwipe®. Measure the absorbance of the sample.

### Problems with Use of Spectrophotometers

If an instrument is incorrectly calibrated or blanked, inaccurate results could occur. Other sources of error include scratched or smudged cuvettes, insufficient volumes, undissolved particles in solution, samples with incorrect concentration, use of incorrect cuvettes or solvent. Slight damage to cuvettes (i.e., scratches, dirty surface) or particulate matter in a sample can cause light scattering rather than light passing through the sample. The resulting absorbance will be higher than the actual value because the instrument will "sense" less light passing through the sample. If an insufficient volume is used in a cuvette, the light will pass over the sample and not be absorbed, resulting in a lower absorbance reading. Cuvettes and solvent should be selected so that they do not absorb light in the range of interest. Glass cuvettes begin to absorb light around 350 nm and continue to do so at lower wavelengths; therefore, quartz cuvettes should be used for UV measurements. If the analyte absorbs in the same region as the cuvette or solvent, the true absorbance will not be recorded.

### Absorbing Species

A number of organic chromophoric groups and inorganic species absorb light in the ultraviolet-visible region. As indicated in Figure 3.1, light absorbed in the UV-Vis region of the electromagnetic spectrum is due to excitation of valence electrons from a lower to a higher energy level. Because of this excitation, absorbing species can be determined directly; however, nonabsorbing species are detected using color-forming reactions. Nonabsorbing species often react with reagents to create colored complexes or molecules that can then absorb.

The absorption of ultraviolet or visible radiation results in electronic transitions involving 1) $\pi$, $\sigma$, and $n$ electrons, 2) d and f electrons, and 3) charge transfer electrons. Each of these electronic transitions will be briefly discussed below. For more in-depth discussions, consult references on spectroscopy (see References and Further Readings).

Covalent bonds can be described using several theories, including the linear combination of atomic orbitals called molecular orbitals. Atomic orbitals are descriptions of electrons in an atom, and molecular orbitals can be viewed as the description of the electrons in a molecule. The molecular orbitals associated with a single bond are designated as sigma ($\sigma$) molecular orbitals and the electrons in those orbitals are called sigma electrons. Multiple bonds contain two types of molecular orbitals; $\sigma$ and pi ($\pi$) molecular orbitals. The nonconstructive combination of the orbitals, are referred to as antibonding molecular orbitals, $\sigma^*$ and $\pi^*$, respectively. In addition to $\sigma$ and $\pi$ electrons, molecules frequently contain unshared electrons called nonbonding electrons (n). Figure 3.3 shows the relative energies of the molecular orbitals.

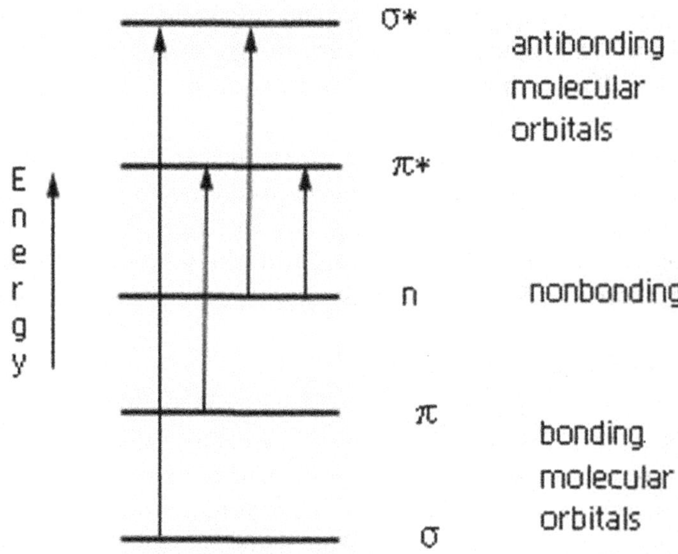



Figure 3.3 – Molecular orbital transitions that can occur when matter interacts
with electromagnetic radiation.

Organic compounds and some inorganic anions are absorbing species containing $\pi$, $\sigma$, and *n* electrons. These species absorb radiation through transitions between molecular orbitals, including $\sigma \rightarrow \sigma^*$, $\pi \rightarrow \pi^*$, $n \rightarrow \sigma^*$, and $n \rightarrow \pi^*$ (see Figure 3.3). Recall that the energy of a photon is related to frequency ($E = h\upsilon$). The frequency of radiation absorbed during the electronic transitions between the molecular orbitals is equal to the difference in energy of the molecular orbitals. A $\sigma \rightarrow \sigma^*$ transition is associated with a large energy difference and absorbs light beyond the ultraviolet-visible range (<150 nm). Therefore, these transitions will not be illustrated in this section of the text. The other transitions ($\pi \rightarrow \pi^*$, $n \rightarrow \sigma^*$, and $n \rightarrow \pi^*$) occur in the UV-Visible region, between 150 and 700 nm. Table 3.1 and Figure 3.4 show examples of biologically important compounds that absorb in the UV-Visible region. Absorptions with molar absorptivity coefficients of less than $10^3$ are considered low intensity and result from low probability (so-called, forbidden) transitions. Molar absorptivity coefficients will be used in many figures in this text because they provide an absolute scale to compare spectral intensities, rather than absorbance, which is dependent on the concentration of the solution utilized.

Proteins contain several chromophores that absorb light in the UV-Vis range. The peptide bond undergoes a $\pi \rightarrow \pi^*$ transition with a $\lambda_{max}$ around 190 nm and $\varepsilon_{max}$ of 7000. A weaker $n \rightarrow \pi^*$ transition occurs between 210-220 nm ($\varepsilon_{max} \sim 100$). The side chains of several amino acids absorb in the UV range, frequently in the same region as the peptide bond (see Table 3.1 and Figure 3.4). Absorption between 190 and 230 nm can often be used to determine the concentrations of peptides or proteins; however, buffer components often also absorb in this region. The amino acids with aromatic side chains (Phe, Tyr, and Trp) absorb light between 250 and 280 nm. The absorption of light by proteins at 280 nm is mainly due to the presence of these residues in the macromolecule (see Figure 3.4).

Table 3.1 – Absorption Characteristics of Some Common Chromophores in Biochemistry.

| Molecule | Structure | Solvent | $\lambda_{max}$ (nm) | $\varepsilon_{max}$ $(M^{-1}cm^{-1})$ |
|---|---|---|---|---|
| Water | $H_2O$ | N/A | 167 | 7000 |
| Nitrate | $NO_3^-$ | Water | 301 200 | 4.2 8900 |
| Acetone | 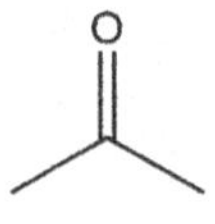 | Water | 19 265 | 35 17 |
| Cysteine | 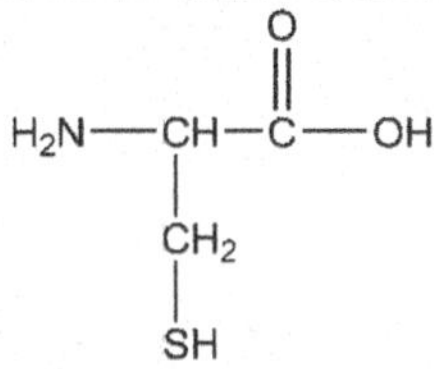 | Water | 23 | 3200 |
| Phenylalanine | 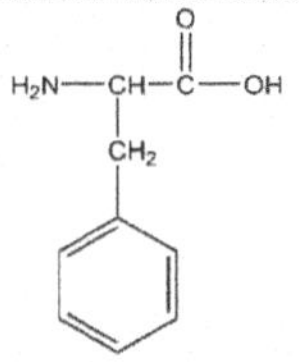 | Water | 257 | 170 |
| Tyrosine | 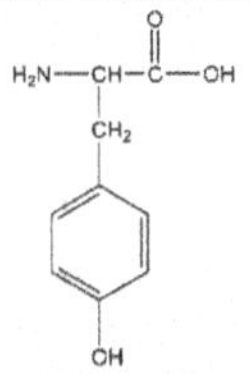 | Water | 223 275 | 7500 1300 |
| Tryptophan | 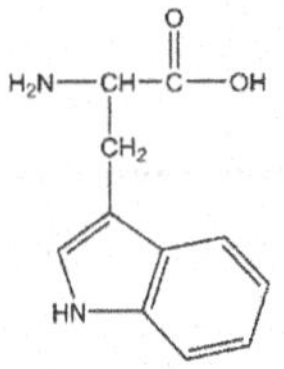 | Water | 218 278 | 33000 5500 |
| Histidine | 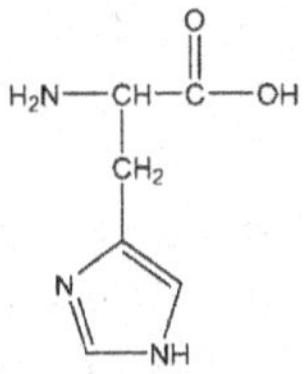 | Water | 211 | 6500 |

## (a) Absorption Spectra for Various Amino Acids

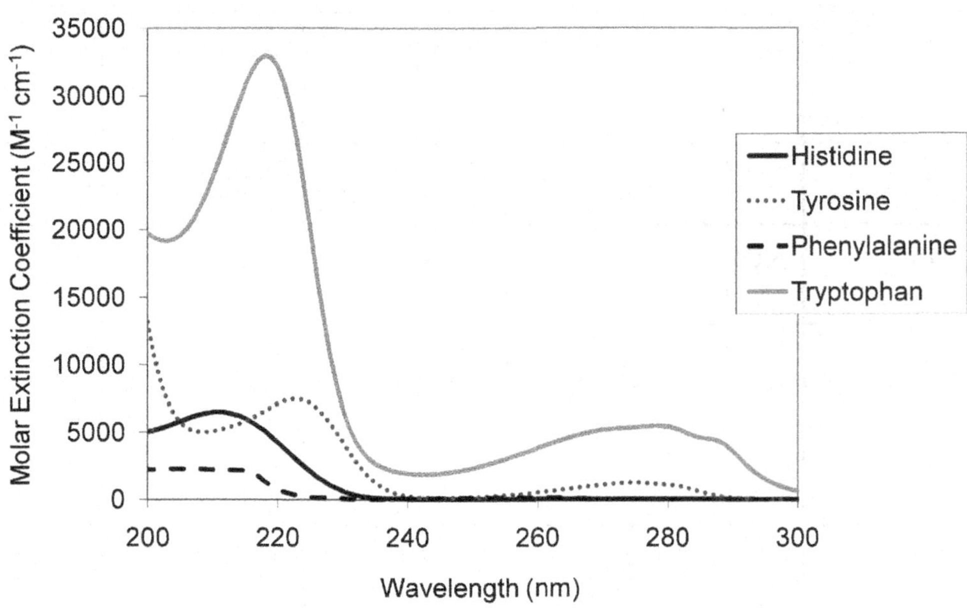


## (b) Absorption Spectra for Various Amino Acids

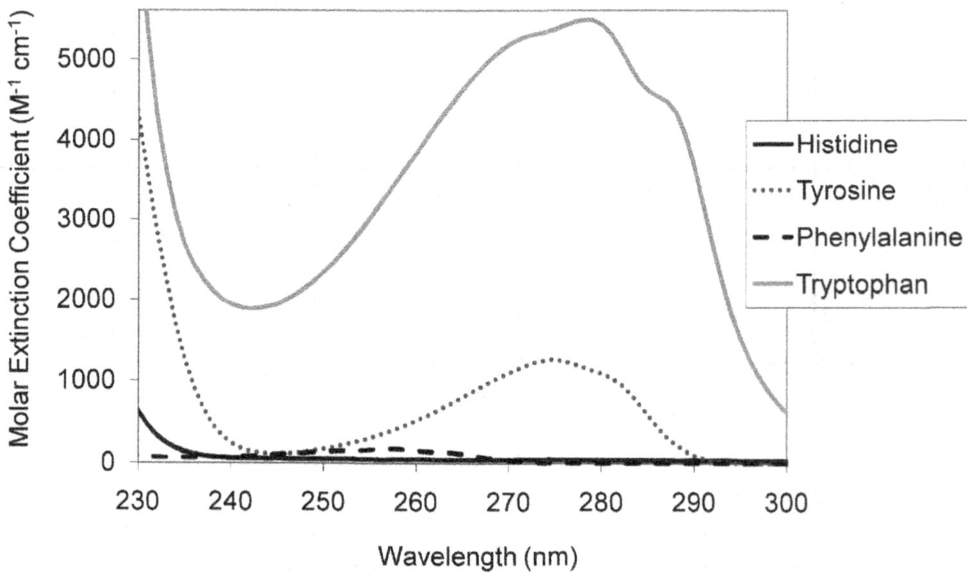


Figure 3.4 – Absorption spectra of several amino acids in water.

If the extinction coefficient of the peptide/protein is known, the concentration of the protein can be derived from the absorbance at 280 nm. If the molar extinction coefficient is unknown and the protein does not contain prosthetic groups or cofactors that absorb at 280 nm, the concentration of the protein can be estimated based on the following formula:

$$\varepsilon_{\text{protein at 280 nm}} = n_{\text{Trp}}\varepsilon_{\text{Trp at 280 nm}} + n_{\text{Tyr}}\varepsilon_{\text{Tyr at 280 nm}} + n_{\text{Cystine}}\varepsilon_{\text{Cystine at 280 nm}},$$

where   $n$ = number of amino acid residues
       $\varepsilon$ = molar extinction coefficient at specified wavelength

The molar extinction coefficients at 280 nm listed in the literature vary slightly (see Table 3.2 for literature values and references).

Table 3.2 – Literature Values of Molar Extinction Coefficients of Chromophores at 280 nm.

| | Tryptophan | Tyrosine | Cystine | Reference |
|---|---|---|---|---|
| Molar extinction coefficient ($M^{-1}cm^{-1}$) | 5,690 | 1,280 | 125 | Ramos, Edelhoch |
| | 5,500 | 1,490 | 125 | Pace, et al. |

Nucleic acids and nucleotides also absorb light in the UV-Vis region due to the presence of aromatic purine and pyrimidine bases. Several different $\pi \to \pi^*$ and $n \to \pi^*$ transitions occur in the region between 200 and 300 nm. Each nitrogenous base has a distinct absorption spectrum (see Table 3.3 for absorption maxima). Usually in nucleic acids, the spectral peaks of the discrete bases merge into a single band with a $\lambda_{max}$ at 260 nm.

Transition metal ions absorb light in the ultraviolet-visible region of the spectrum through electronic transitions involving the d and f electrons. The absorbance of a metal ion cofactor can also be used to quantify the amount of protein in the sample or investigate the oxidation state of the cofactor. Charge-transfer absorption occurs commonly in inorganic complexes and some organic complexes. This absorption results from the transfer of an electron from an electron donor to an electron acceptor. An example of this transition is the transfer of an electron from a ligand, such as thiocyanate, to a transition metal ion, such as $Fe^{3+}$.

### *Effect of Solvent on Absorption*

As mentioned previously, substances frequently absorb light in the UV-Vis region of the absorption spectra. Choice of solvent, therefore, should include a consideration about the transparency of the solvent in the region of the spectrum to be investigated. The other consideration is the effect of the solvent on the absorbing species. The local environment of a molecule can affect its energies in the ground and excited states. A more polar solvent will stabilize all electronic states of a molecule and will tend to remove spectral details due to vibrational effects. Also the polarity of the solvent will influence the absorption maxima of the absorbing species. Polar solvents interact with molecular orbitals as follows $n >> \pi^* >> \pi$. (See Figure 3.3.) The $\pi \to \pi^*$ transition shifts to a lower energy (red-shift) due to the increased stabilization of the $\pi^*$ orbital relative to the $\pi$ orbital. The $n \to \pi^*$ transition would undergo a blue-shift (shift to higher energy) because of the stabilization of the n orbital relative to the $\pi^*$ orbital. The most dramatic shifts will occur in systems in which the strongest interactions between the molecule and solvent occur (i.e., a polar chromophore with a polar solvent). Figure 3.5 shows the effects of solvent on the absorption spectra of acetone. As shown in the figure, the $\pi \to \pi^*$ transition changes as a result of solvent polarity. In a more polar solvent, such as water, the absorption maximum occurs at a shorter wavelength (higher energy) as compared to a nonpolar solvent, such as hexane.

Table 3.3 – Absorption Maxima for Nucleotides

| Nucleotide | Structure of base | $\lambda_{max}$ (nm) | $\varepsilon_{max}$ at 260 nm |
|---|---|---|---|
| AMP | 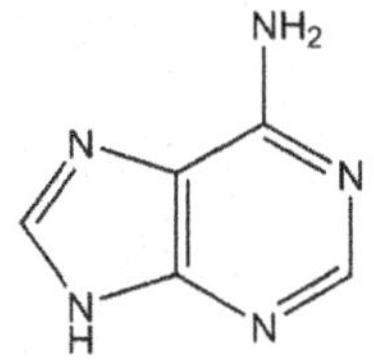 | 260 | 15,400 |
| CMP | 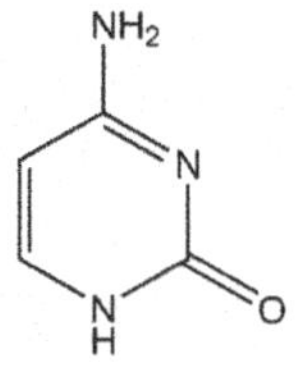 | 272 | 7,500 |
| GMP | 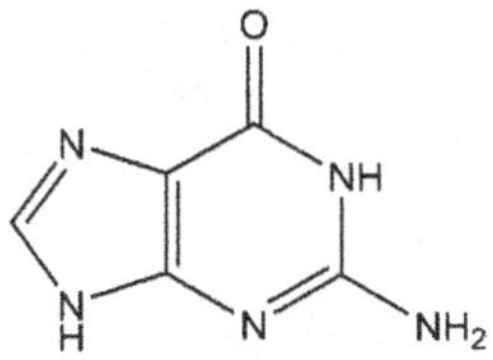 | 250 | 11,700 |
| dTMP | 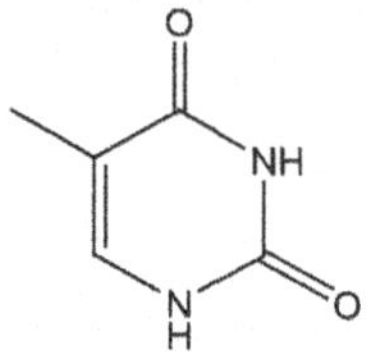 | 268 | 9,200 |
| UMP | 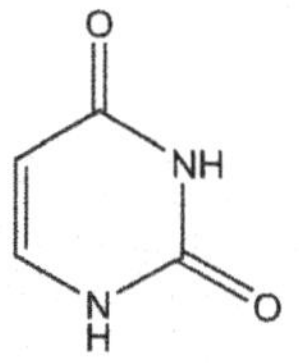 | 260 | 9,900 |

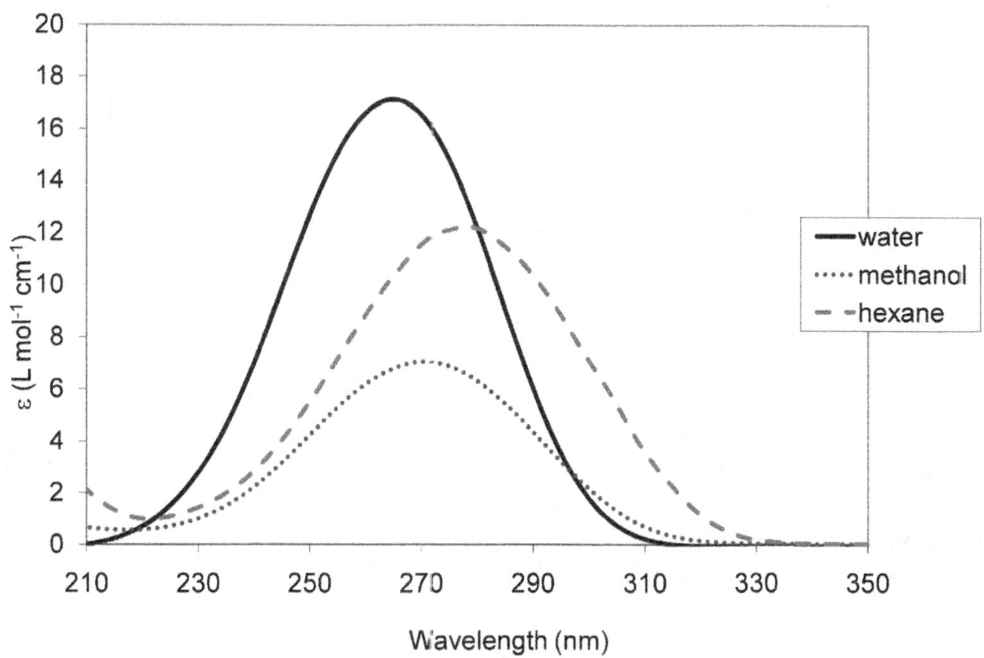


Figure 3.5 – Effects of solvent on the absorption spectra of acetone.

The absorption spectra of biological molecules, such as Tyr and Trp, are also affected by the solvent. The topology of a peptide/protein can be studied by looking at the effect of solvent on the absorbance at 280 nm. A similar effect is observed in the local environment of a protein. Tryptophan residues that are buried in a hydrophobic environment in a protein will exhibit a spectral shift. Scientists will often study the denaturation of protein by examining changes in wavelength of maximum absorbance as the overall protein structure changes.

## *Chemical Effects on Absorption*

The absorption of some chromophores can be affected by chemical changes, such as a change in pH or ionic strength and changes in dipole coupling. The absorption of chromophores in peptides, proteins, and DNA depends on the structure of the macromolecule. The denaturation of these structures results in changes in the dipolar couplings, which result in a change in the absorption of light. The denaturation of proteins and nucleic acids can therefore be monitored or measured using UV absorption spectroscopy. Other methods for monitoring protein denaturation include circular dichroism, intrinsic fluorescence, and size-exclusion chromatography (see References and Further Reading).

The effect of pH on the absorption of tyrosine and phenylalanine is shown in Figure 3.6. The side chain of phenylalanine is unaffected by change in pH, but the phenolic OH on tyrosine changes ionization state as pH changes (see Figure 3.6). This change in the electronic structure causes a large shift in the absorption spectrum (see Figure 3.7).

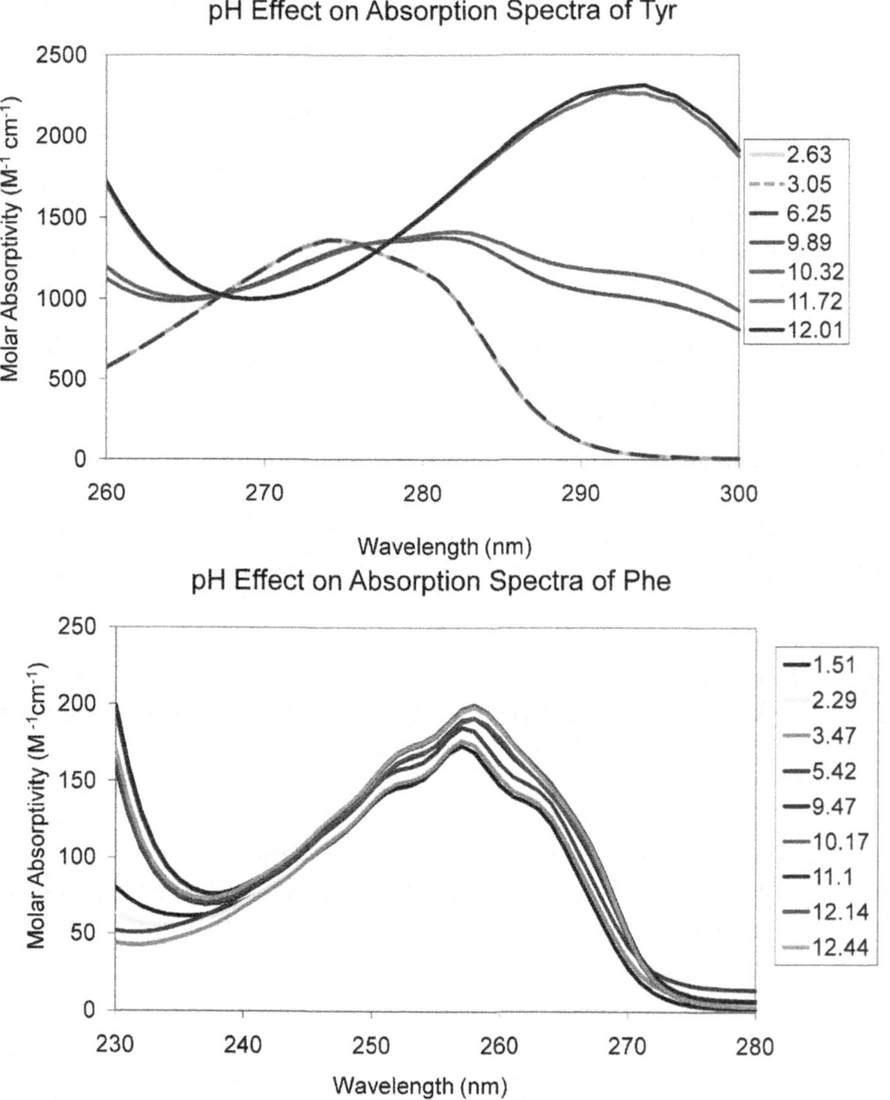

Figure 3.6 – Structure of tyrosine side chain under acidic and basic conditions.

The ionic strength of a solution can affect the structure of biological molecules by altering the intermolecular interactions that influence the overall structure of the macromolecule. For example, in high ionic strength solutions the base pairing interactions in nucleic acids are weakened, and this can lead to shifts in the absorbance peaks for the DNA. Unless spectral shifts are important for study, it is important that the same solvent be used in a given spectroscopy study.  Also conditions such as pH and ionic strength should also be maintained.

Figure 3.7 - The effect of pH on the absorption of tyrosine and phenylalanine. Note in the spectra of tyrosine, the curves for the pH at 2.63 and 3.05 are aligned with the curve for pH 6.25.

## 3.3 MOLECULAR FLUORESCENCE SPECTROSCOPY

Molecular fluorescence spectroscopy, phosphorescence spectroscopy, and chemiluminescence spectroscopy are three related techniques collectively called molecular luminescence spectrometry. Luminescence is the emission of light by a substance that occurs when an electron returns to its ground state from some excited state. This textbook will only focus on fluorescence spectroscopy. If you are interested in learning more about the other forms of luminescence spectrometry, please see References and Additional Reading.

The simplest form of fluorescence is resonance radiation or resonance fluorescence. This phenomenon is observed in gas samples and dilute atomic vapors. The sample absorbs light of a specific wavelength ($\lambda_{abs}$), and the electrons are excited. After a brief period of time (approximately $10^{-7}$ s), the electrons return to the ground state and emit radiation of the same wavelength (i.e., $\lambda_{emitted} = \lambda_{abs}$).

Many molecules exhibit resonance fluorescence; however, the fluorescence normally observed is ordinary fluorescence, in which a molecule emits light of a longer wavelength (lower frequency) than absorbed. Therefore, two properties dictate the fluorescence process and are specific for molecules: maximum excitation wavelength and maximum emission wavelength. A molecule absorbs UV radiation of an appropriate wavelength, and electrons are promoted from a vibrational level in the ground state to one of many vibrational levels in the excited state, usually the first excited singlet state. The electron will quickly fall to the lowest vibrational level of the excited state by losing energy through molecular collisions. This is referred to as radiationless transfer of energy or vibrational relaxation and is shown with the zigzag arrow in Figure 3.8. As the electrons return to the ground state (from the lowest vibrational level of the excited state), energy is emitted in the form of light, which is referred to as fluorescence. As shown in Figure 3.8, the emitted light is a longer wavelength than the absorbed light due to energy loss by the molecule prior to emission. Figure 3.9 shows an example of an excitation and emission spectra for a Hoescht 33258.

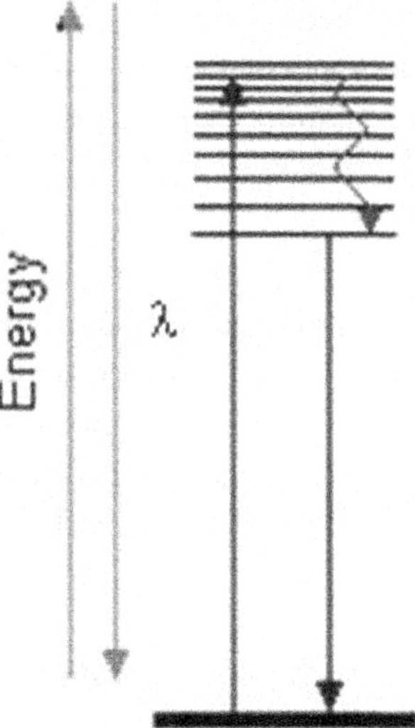


Figure 3.8 – Electron transitions occurring in fluorescence.

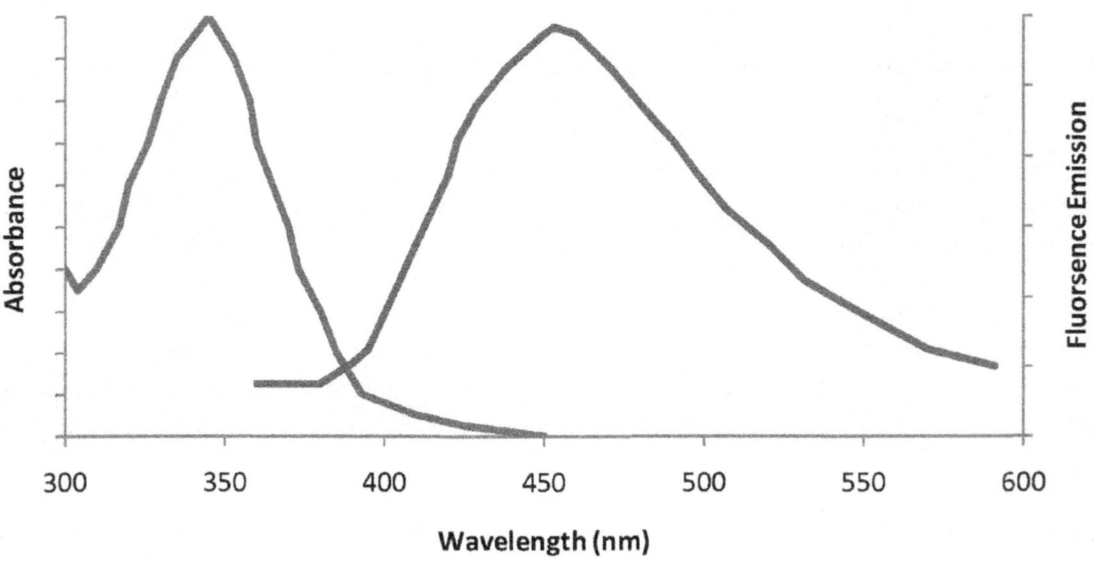


Figure 3.9 – Excitation and emission spectra of Hoechst 33258. Excitation maximum occurs at 350 nm; emission maximum at 450 nm.

### *Fluorescing Species*

Fluorescence is most commonly associated with $\pi \to \pi^*$ transitions rather than $n \to \pi^*$ transitions because of quantum efficiency. Quantum efficiency is the ratio of the number of molecules that fluoresce versus the total number of excited molecules. It is less likely that other competing processes will occur with $\pi \to \pi^*$, and fluorescence will be observed.

Fluorescence is most frequently observed in molecules with fused aromatic ring systems and monosubstituted aromatic ring systems.  Certain substitutions on or within the ring system can cause a decrease in fluorescence. Simple heterocyclic compounds such as furan and pyrrole do not fluoresce, but, if these rings are fused to a benzene ring, fluorescence is seen. Heavier atom halogens (bromine and iodine) decrease the fluorescence of the species as do carbonyl or carboxylic functional groups.

Temperature, solvent, and pH can affect fluorescence. Increasing the temperature causes an increase in the frequency of collisions between molecules, which can lead to radiationless transfer of energy between the chromophore and itself or some other molecule, thus leading to a decrease in fluorescence. Solvents, especially those with heavy atoms, such as bromine or iodine, have a similar effect as substitution with these atoms. For acidic or basic compounds, a change in the pH will change the electronic structure of the compound, and this can result in a shift in the excitation or emission maxima.

In principle, molecular fluorescence spectrometry is more sensitive and specific than absorption spectrometry because fluorescence has zero background. Fluorometers are more sensitive than spectrophotometers by 3 to 6 orders of magnitude. Fluorometers are also less sensitive to interferences because fewer molecules absorb light and also emit light, especially of the same wavelength as the target emission wavelength. However, these techniques are often less applicable than UV-Vis spectroscopy because more molecules absorb UV-Vis radiation than exhibit photoluminescence.

### *Two Primary Types of Instruments*

There are two types of spectrofluorometers available in laboratories. Undergraduate laboratories most frequently contain filter fluorometers, mainly due to their lower cost.

1) Filter fluorometer – measures the amount of light emitted at a single wavelength when the sample is irradiated at one wavelength. These instruments are a good choice when sensitive quantitative measurements for specific compounds are desired. These instruments are also lower in cost than spectrofluorometers.

   The specific wavelengths are provided by optical filters. A light source in a broad excitation wavelength range for the compound passes through the filter, which blocks out other wavelengths. The specific radiation passes through the sample and excites it. Fluorescence occurs in all directions, but the fluorescence is detected at right angles to the excitation beam to avoid light scattering from the excitation source. The emission light passes through a second filter and is observed by the detector.

2) Spectrofluorometer – uses excitation and emission monochromators in place of the filters. These instruments allow for varying wavelength selection. The instrument can scan over a range of wavelengths and can provide both excitation and emission spectra.

# 3.4 INFRARED SPECTROSCOPY

Infrared spectroscopy (IR) is used primarily to identify the presence or absence of discrete functional groups in a molecule. Infrared radiation is lower energy radiation than UV-Vis (see Figure 3.1); therefore, it does not have enough energy to induce electronic transitions. Molecules absorb IR radiation through changes in vibrational and rotational states. In liquids or solids, rotational transitions are not discrete and are less useful for spectroscopic studies; therefore, this discussion will focus only on molecular vibrations.

The atoms in a molecule are not fixed in space and can move due to a number of different vibrations. The two main types of vibrations are stretches and bends. A stretch occurs when the interatomic distance changes along the bond axis either symmetrically or asymmetrically. A bend occurs when the angle between two bonds is changed. There are fours main types of bends: rocking, scissoring, wagging, and twisting.

A molecule will absorb IR radiation if the frequency of the radiation matches the vibrational frequency of the molecule. Absorption of the light will increase the amplitude of the molecular vibration. Some molecules exhibit vibrations but do not absorb IR radiation. In order for a molecule to absorb IR, the vibrations must cause a net change in the dipole moment; for example, if a homonuclear diatomic molecule is linear and stretches symmetrically, no change in the dipole will occur, and this molecule will not absorb IR radiation. Molecules such as these are said to be IR inactive.

The frequency of vibration depends on two factors: the mass of the atoms and the strength of the bond. Heavier atoms will vibrate more slowly than lighter ones; therefore, the frequency of bond vibration decreases with increasing atomic mass. Stronger bonds are usually stiffer, requiring more force to stretch or compress them; thus, stronger bonds vibrate faster than weaker bonds (assuming similar masses), and the frequency of vibration increases with bond energy. Table 3.4 shows characteristic infrared absorptions for some common bond types.

Table 3.4 – Characteristic Infrared Absorptions.

| Bond | Frequency range (cm$^{-1}$) |
| --- | --- |
| alkyl C-H | 2853-2962 |
| alcohol O-H | 3590-3650 |
| amine N-H | 3300-3500 |
| C≡C | 2100-2260 |
| C≡N | 2220-2260 |
| C=C | 1620-1680 |
| C=O | 1630-1780 |

IR spectroscopy is frequently used by organic and inorganic chemists to identify structural features in molecules. It is used less frequently in biochemical application because of the large size and complexity of many macromolecules. Some clinical chemists have demonstrated that IR can be used to measure small molecules in biological samples such as plasma (see References and Further Reading).

## 3.5 NUCLEAR MAGNETIC RESONANCE SPECTROSCOPY

Nuclear Magnetic Resonance (NMR) spectroscopy is a powerful analytical tool, especially for scientists interested in determining information about the molecular structure of a compound. NMR spectroscopy involves the interaction of radio waves with a molecule; the interaction of these low-energy radio waves with a molecule causes nuclear spin flips. The chemical environment of specific nuclei will be studied by this technology.

Subatomic particles such as electrons, protons, and neutrons can be viewed as spinning tops. Each of these particles is assigned quantum numbers depending on the direction of the spin. Table 3.5 shows the nuclear spin quantum numbers of nuclei. In many atoms (for example $^{12}$C), the spins of the nuclear particles (protons and neutrons) cancel each other out such that the nucleus has no overall spin. Nuclei that do not spin will not give rise to an NMR spectrum.

Table 3.5 – Nuclear Spin Quantum Numbers for Nuclei.

| Mass Number | Atomic Number | Spin Quantum Number, I |
| --- | --- | --- |
| Even | Even | 0 |
| Even | Odd | 1,2,3 … |
| Odd | Odd or Even | 1/2, 3/2, 5/2,…. |

The most common nucleus with a spin studied by NMR is a proton. Other forms of NMR spectroscopy utilize $^{13}$C, $^{14}$N, and $^{31}$P nuclei. It is important to know the spin quantum number because it reveals the number of possible orientations for the nuclei. One can determine the number of orientations that the spinning nuclei can assume in an external magnetic field using the formula 2I +1. For example, protons have I = ½ and can exist in two energy levels (i.e., 2(½) +1): one with spin -½ (higher energy) and one spin +½ (lower energy). (See Figure 3.10.)

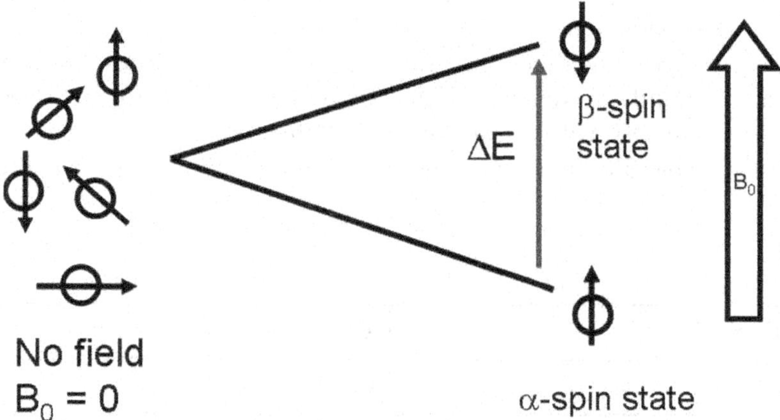


Figure 3.10 – Spin states of protons in the absence ($B_0 = 0$)
and presence of an external magnetic field ($B_0$).

The strength of the magnetic field affects the energy difference between the two spin states. In a strong magnetic field, the energy difference between the two spin states is greater than in a weaker field. The energy difference is related to the strength of the magnetic field, Planck's constant, and the gyromagnetic ratio ($\gamma$) and can be calculated using the following formula:

$$\Delta E = h\nu = (h/2\pi)\gamma B_0,$$

where    $\gamma$ is a constant that depends on the magnetic moment of the nuclei under study
for a proton, $\gamma$ is 26,753 $sec^{-1}$ $gauss^{-1}$

h    = Planck's constant = $6.626 \times 10^{-34}$ J·s

$B_0$ = strength of magnetic field

The energy difference between spin states is not that large for a magnetic field of 2.5 T (25,000 gauss); it is only $4 \times 10^{-5}$ kJ/mol. NMR spectrometers, however, can detect these small differences in energy. When the nuclei are in a magnetic field, the initial population of the nuclei can be determined based on the Boltzmann distribution. Accordingly, the lower energy level will contain a large population of nuclei.  It is possible to promote nuclei to the higher spin state by exposing the protons to electromagnetic radiation. When a proton interacts with a radio-frequency photon that has the same energy as the energy difference between the spin states, the nuclei can flip spin. This spin flip is often referred to as resonance. The term nuclear magnetic resonance spectrometry is derived from the fact that a nucleus is undergoing a resonance transition in the presence of a magnetic field.

The magnetic field experienced by the nucleus is not always equal to the strength of the external magnetic field. Electrons surrounding the nucleus often shield the nucleus from the applied magnetic field, and, therefore, the nucleus "feels" a lower magnetic field. This shielding decreases the energy difference between the spin states relative to the energy difference between the spin states if the nucleus was not surrounded by electrons. In order to achieve resonance, a higher external magnetic field must be applied to the sample.  In other cases, often observed with pi electrons, the electron spin may enhance the magnetic field experienced by the nucleus. These protons are deshielded and require less of an applied magnetic field to achieve resonance.

In an NMR spectrometer, the frequency of the radio waves is constant. Depending on the electronic environment of the nucleus, the energy difference between the spin states will vary. The external magnetic field is adjusted until the energy difference between the spin states equals the energy of a photon of radiation, and radiation is absorbed. A chemical shift is defined as the amount of applied magnetic field necessary to achieve resonance. Due to the shielding effects of electrons, the chemical shift is a function of the nucleus and its chemical and electronic environment. Typical spectrometers will measure chemical shift relative to a standard such as tetramethylsilane (TMS). Table 3.6 lists some common chemical shifts for various protons in compounds.

Table 3.6 – Common Proton Chemical Shifts.

| Type of Hydrogen | Chemical Shift, $\delta$ (ppm) |
|---|---|
| Aliphatic hydrogens | 0.8-1.7 |
| Allylic hydrogens | 1.6-1.9 |
| Benzylic hydrogens | 2.2-2.5 |
| Acetylenic hydrogens | 2.5-3.1 |
| Vinyl hydrogens | 4.6-5.7 |
| Aromatic hydrogens | 6.0-8.5 |
| Amine hydrogens | 1.0-5.0 |
| Alcohol hydrogens | 0.5-6.0 |
| Hydrogens on carbons adjacent to oxygen | 3.3-4.0 |
| Hydrogens on carbons adjacent to carbonyl | 2.3-3.0 |
| Carboxyl hydroxyl hydrogens | 10.0-13.0 |

### Spin-spin Coupling of Protons

The peaks resulting from absorption of radio waves are often split into multiple peaks. These peak splittings often are recognizable patterns, such as doublets, triplets, quartets. This spin-spin splitting or coupling occurs when a proton "senses" the number of equivalent protons on the carbon next to it. The spin of a neighboring proton influences the magnetic field experienced by the nucleus, similar to the effect of electrons. Analysis of the splitting pattern can reveal information about the number of protons on adjacent carbons.

NMR spectrometry is increasingly being used in biochemistry to elucidate the structures of numerous biomolecules (see References and Further Readings). Some undergraduate institutions may not have a high-field instrument immediately available to complete these experiments in a hands-on fashion, but biochemistry students should be aware of the utility of this spectroscopic technique.

### Chemical Shifts of Carbon-13

The above discussion of NMR focused on protons. The chemical shifts of carbon-13 atoms follow similar characteristics to those mentioned for protons. Table 3.7 lists common chemical shifts for carbon-13 atoms.

Table 3.7 – Common Carbon-13 Chemical Shifts

| Type of Carbon | Chemical Shift, $\delta$ (ppm) |
| --- | --- |
| Primary alkyl | 0 – 40 |
| Secondary alkyl | 10 – 50 |
| Tertiary alkyl | 15 – 50 |
| Alkyl halide or amine | 10 – 65 |
| Alcohol or ether | 50 – 90 |
| Alkyne | 60 – 90 |
| Alkene | 100 – 170 |
| Aryl | 100 – 170 |
| Nitrile | 120 – 130 |
| Amides | 150 -180 |
| Carboxylic acids, esters | 160 – 185 |
| Aldehydes, ketones | 182 – 215 |

## *Two-Dimensional NMR Spectroscopy*

To this point, only one-dimensional NMR has been described, in which the data are frequently plotted as intensity versus chemical shift. In a two-dimensional experiment, the intensity is plotted as a function of two frequencies or chemical shifts. There are numerous different types of two-dimensional NMR spectroscopies, but the most common forms are H-H Correlation Spectroscopy (COSY) and Heteronuclear Correlation Spectroscopy (HETCOR). In COSY, the proton spectrum of the compound is plotted on both axes and the resulting contour plot is a correlation of the proton resonances. The proton spectra are plotted on the axes, and the correlation contour plot is shown. The significant information that is discerned about the molecule is determined from the peaks appearing off of the central diagonal line. These correlations indicate information about neighboring protons.

HETCOR involves correlating spectra of two different nuclei (frequently protons with attached heteronuclei such as carbon-13). Other correlation spectroscopies, such as Heteronuclear Multiple Quantum Coherence (HMQC) and Heteronuclear Single Coherence (HSQC), have been developed that correlate attached heteronuclei in more sensitive manners. Biological NMR involves using various correlation NMR experiments to probe relationships between the nuclei ($^{15}$N and $^{13}$C) in the molecules. If you are interested in learning more about two-dimensional spectroscopy, please see References and Further Readings.

Science in Action
***Box 3.2 – Magnetic Resonance Imaging (MRI)***

Magnetic Resonance Imaging is an important application of NMR for the medical field. The technique emerged in the early 1970s to mid-1980s. In 2003, Pail C. Lauterbur and Sir Peter Mansfield received the Nobel Prize in Medicine for their contributions to the development of MRI. An MRI instrument measures the time it takes for a proton to relax after a spin flip has occurred. The relaxation time depends on the environment in which the proton is located. For example, the protons in water bound to a macromolecule will relax faster than those in free fluid. Also, the tissue or structure in which the water is associated will alter the relaxation times. These NMR data can be converted to three-dimensional images of the interior of the objects. MRI is a noninvasive technique that enables physicians to see the interior of humans without potential injuries that result from techniques such as X-rays.

## 3.6 MOLECULAR MASS SPECTROMETRY

Mass spectrometry (MS) is an analytical technique that can be used to determine the elemental composition of a molecule, elucidate the structure of the molecule, and evaluate the qualitative and quantitative composition of mixtures. It is frequently used in biochemistry to determine the structure of macromolecules such as proteins and polysaccharides. One of the most common exposures for undergraduate students to MS is as a detector following chromatographic methods, such as Gas Chromatography and Liquid Chromatography (see Chapter 7).

Mass spectrometers separate ions based on their mass (m) to charge (z) ratio (m/z). The instrument counts the number of ions of each type and reports the abundance of those ions. The first step in analysis involves converting neutral molecules into molecular ions by an ionization technique, such as bombardment with electrons (e.g., electron impact, EI). These ions are less stable and are often in the excited state following the ionization process. As the excited ions return to the ground state, fragmentation often occurs to generate ions of lower molecular weight. Molecules will fragment in predicable patterns, such as through loss of $CH_3$ or $H_2O$. The mixture of ions is separated based on m/z ratio and quantified by the detector. Depending on the ion source (ionization agent), few or many fragmentations will occur. Electron impact generally produces extensive fragmentation. Softer modes of ionization include chemical ionizations, plasma desorption ionization, fast atom bombardment ionization, electrospray ionization, and matrix-assisted laser desorption ionization (MALDI).

Several mass analyzers can be used to mass-analyze the ions; a few of the ones undergraduates may encounter are described below. A scanning mass analyzer uses electromagnetic fields to separate the ions according to their m/z ratio and a slit to select ions with specific m/z ratio that reach the detector. The different m/z ratios are then scanned across the detector slit, and the ion current is recorded. In a magnetic deflection mass spectrometer, the ions are accelerated to a high velocity, and an applied magnetic field separates the ions according to the m/z ratio. A quadrupole mass analyzer works like a mass filter. Potentials are applied to the quadrupole rods that cause selected m/z ratios to pass through. Ions that do not have a straight trajectory through the quadropole collide with the quadropole rods and do not reach the detector. A time of flight mass spectrometer measures the time it takes ions of different masses to move from the ion source to the detector.

This requires that the starting time (the time at which the ions leave the ion source) is well-defined. More details about mass analyzers can be found in References and Further Readings.

## *MALDI-TOF and Protein Mapping*

One advance in the area of biological MS has been the development of MALDI-TOF. MALDI is a soft ionization method and is able to ionize and analyze large-molecular-weight compounds (up to 500,000 Da). A sample is combined with a matrix material and applied to a plate. The sample is dried and then exposed to a laser. The sample is volatilized and proton transfer occurs from the matrix to the sample molecule, resulting in positively charged ions. The resulting ions exhibit minimal fragmentation and are then separated by TOF.

MALDI-TOF is frequently used for protein mapping. A protein or peptide sample is digested with an enzyme, frequently trypsin. The sample of digestion fragments is then analyzed using MALDI-TOF. The resulting spectrum is often considered a "fingerprint" for the protein and is searched against a database to determine the identity of the protein.

If the protein sequence is not determined in this fashion, tandem mass spectrometry methods are often used to identify the protein sequence. Tandem mass spectrometry involves connecting two mass spectrometers in series. The first mass spectrometer separates ions and selects one m/z value. The selected ions enter a collision cell in which they collide with argon and are fragmented. The fragment ions are separated and the m/z values determined in the second mass spectrometer.

## *Interpretation of the Mass Spectrum*

The following list is a set of general guidelines that can assist you in interpreting mass spectra of small molecules produced by methods such as EI:

1) Determine the molecular ion $(M^+)$ or $(M+1)^+$ peak, where M = m/z ratio, equivalent to the molar mass when z = 1. The molecular ion peak is usually the most abundant ion in the high-mass cluster. In Figure 3.11, a peak at 181 m/z is observed for tyrosine. In Figure 3.12, the molecular ion for 2-deoxyadenosine is not observed (the molecular weight is 251 g/mol). Compounds containing only C, H, O, S, Si, or an even number of nitrogens have even molecular weights and even $M^+$ peaks. The $M^+$ peak is odd when a compound contains an odd number of nitrogens. This information can be useful in determining the composition of a molecule. In the case of tyrosine, which contains one nitrogen, the $M^+$ peak is odd. Also, the abundance of the $M^+$ will provide information about the stability of the molecule. For example, aromatic compounds give prominent M+ peaks.

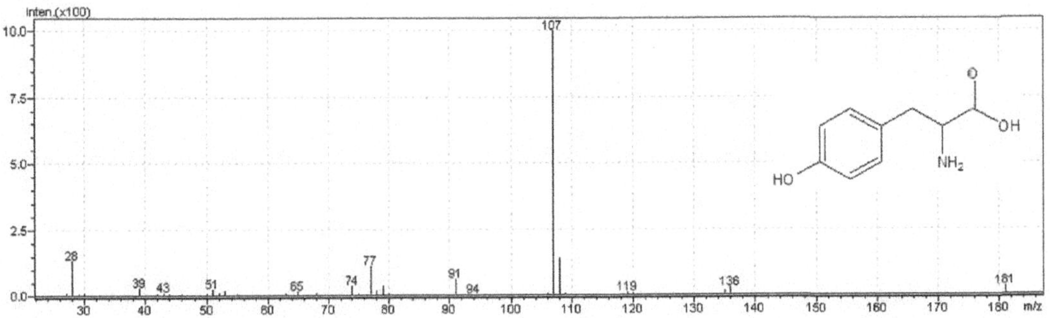

Figure 3.11 – Mass spectrum of tyrosine.

2)  Calculate the elemental composition based on the relative abundances of $M^+$, $(M+1)^+$, and $(M+2)^+$ peaks. The numbers of C and N can be found from the relative ratios of $M^+$ and $(M+1)^+$ peaks, whereas the numbers of O, S, Si, Cl, and Br can be found from the ratios of $M^+$ and $(M+2)^+$ peaks.

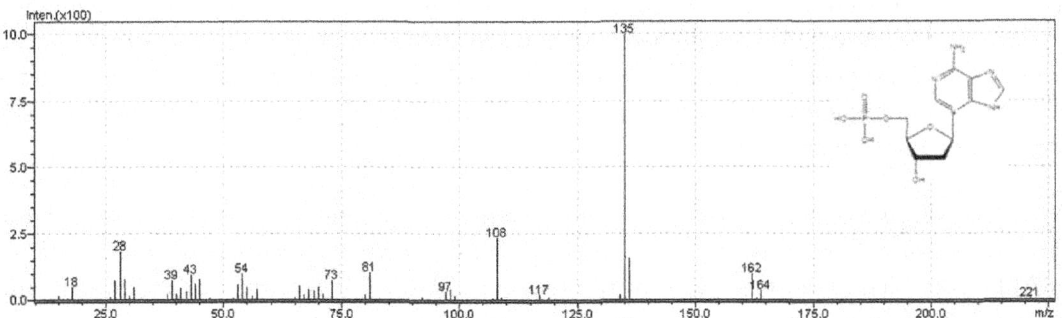

Figure 3.12– Mass spectrum of 2'-deoxyadenosine.

3)  Look for fragmentation patterns by examining the differences in the masses of ion peaks. Common fragmentations will include alpha cleavages, rearrangement reactions, etc. Table 3.8 shows some of the simplest and most common fragments observed in a mass spectrum. There are specific peaks that can be attributed to specific functional groups (i.e., alkanes will contain peaks at 29, 43, 57, and aromatic rings will often have peaks at 39, 51, 65, 77, 91). For more details about fragmentation patterns, see books on mass spectroscopy.

---

## Science in Action
### *Box 3.3 – Proteomics*

Proteomics is a growing scientific field. The proteome is the name given to describe the proteins in the cell. Unlike the genetic material, the proteome is a dynamic substance, constantly changing. Proteomics seeks to identify the proteins in the cell, measure the abundance of proteins in the cells, and identify changes in the proteome with environmental stimuli. This growing field relies on advances in analytical techniques, especially MS, in order to study the amounts and structures of proteins in organisms.

Table 3.8 – Common Mass Spectrum Fragments.

| Commonly Lost Fragments | | Common Stable Ions | |
|---|---|---|---|
| Fragment | Observed change in m/e of the peak | Ion | m/e of the ion |
| $\bullet CH_3$ | m-15 | $CH_3^+$ | 15 |
| $\bullet OH$ | m-17 | $(CH_3CH_2)^+$ | 29 |
| $H_2O$ | m-18 | $(CH_2OH)^+$ | 31 |
| $\bullet CN$ | m-26 | $CH_2=CH\text{-}CH_2^+$ | 41 |
| $H_2C=CH_2$ | m-28 | $H_3C\!-\!C\!\equiv\!\overset{+}{O}$ | 43 |
| $\bullet CH_2CH_3$ or $\bullet CHO$ | m-29 | (benzyl cation: ring–$\overset{+}{C}H_2$) | 91 |
| $\bullet OCH_3$ | m-31 | | |
| $\bullet Cl$ | m-35 | | |
| $H_3C\!-\!\overset{\bullet}{C}\!=\!O$ | m-43 | | |
| $\bullet OCH_2CH_3$ | m-45 | | |
| (benzyl radical: ring–$\overset{\bullet}{C}H_2$) | m-91 | | |

# 3.7 PROTEIN ASSAYS

1) Spectrophotometric Assay

As mentioned above, aromatic amino acids absorb light in the range of 250-280 nm. As the absorption of light by Trp has the highest molar absorptivity coefficient, the measurement of absorbance at 280 nm of peptides/proteins is one of the most common methods to quantitate proteins. It should be noted that this method is very sensitive to differences in amino acid composition and in pH (see Chemical Effects on Absorption under Section 3.2). As shown in Table 3.9, the value of $A_{280}$ will vary for a 1 mg/mL solution of protein, depending on protein composition.

It has been estimated that a 1 mg/mL mixture of proteins will have an absorbance at 280 nm of 1 in a 1 cm cuvette (ref. Bollag). Although this is not entirely accurate, it can be used to approximate the concentration of protein in a sample by measuring the absorbance at 280 nm in a quartz cuvette. To obtain a more accurate concentration, the molar extinction coefficient of the protein must be known.

The advantages of this method are that it is quick and simple, nondestructive, and does not require special reagents. Most buffers used in protein experiments do not absorb around 280 nm, and will not interfere with the measurements. Some disadvantages include moderate sensitivity (50-100 µg of protein detected) and interferences from other macromolecules. Other biochemical molecules from cellular extracts can interfere with the protein determination.

Table 3.9 – Absorbance at 280 nm for 1 mg/mL samples of protein<br>(Adapted from Bollag, D.M. and Edelstein, S.J.).

| Protein | $A_{280}$ (1 mg/mL) |
|---|---|
| Bovine Serum Albumin | 0.70 |
| Ribonuclease A | 0.77 |
| Ovalbumin | 0.79 |
| Enterotoxin | 1.33 |
| γ-Globulin | 1.38 |
| Trypsin | 1.60 |
| Chymotrypsin | 2.02 |
| α-amylase | 2.42 |

Nucleic acids have a maximum absorbance at 260 nm and exhibit significant absorbance at 280 nm. One can calculate the $A_{280}/A_{260}$ ratio to determine the extent of nucleic acid contamination. Pure proteins have a $A_{280}/A_{260}$ ratio of 1.75, whereas nucleic acids have a ratio of approximately 0.5.

This method is often utilized for rapid approximation of protein concentrations or for detecting proteins in fractions following chromatography (see Chapter 7). Quartz cuvettes must be used for this technique, as glass or plastic absorb light at 280 nm.

2) Bradford Assay

Coomassie Brillant Blue G-250 (structure shown in Figure 3.13) is a hydrophobic dye that has four different ionic forms. The absorption peaks for the cationic forms of the dye have maxima around 650 and 470 nm. The anionic form of the dye, which binds preferentially to arginine and lysine residues, has a maximum absorbance to 595 nm. The complexing of the protein to the dye causes a shift in maximum absorbance to 595 nm and a decrease in the absorbance at 465 nm (see Figure 3.14). Therefore, protein concentration can be determined by measuring the absorbance at 595 nm.

This method of protein determination is fairly fast (the samples can be measured 5 minutes after mixing) and sensitive (0 µg/mL – 1500 µg/mL). Certain substances, such as strongly basic buffers, some detergents such as Triton X-100 and SDS, glycerol, acetic acid, ammonium sulfate, and nucleic acids, interfere with this technique. Plastic cuvettes should be used for this technique, as the dye often adheres to and stains glass or quartz cuvettes. The dye can be frequently removed by rinsing with alcohol. Standard curves are prepared with the Bradford assay and bovine serum albumin (BSA) or Immunoglobulin G.

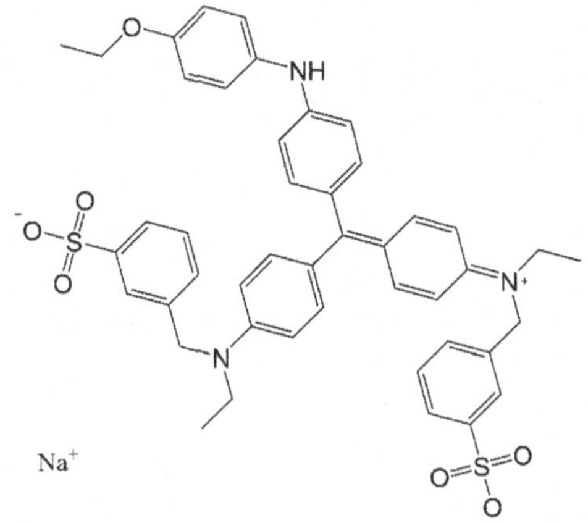

Figure 3.13 – Structure of Coomassie Brillant Blue G-250.

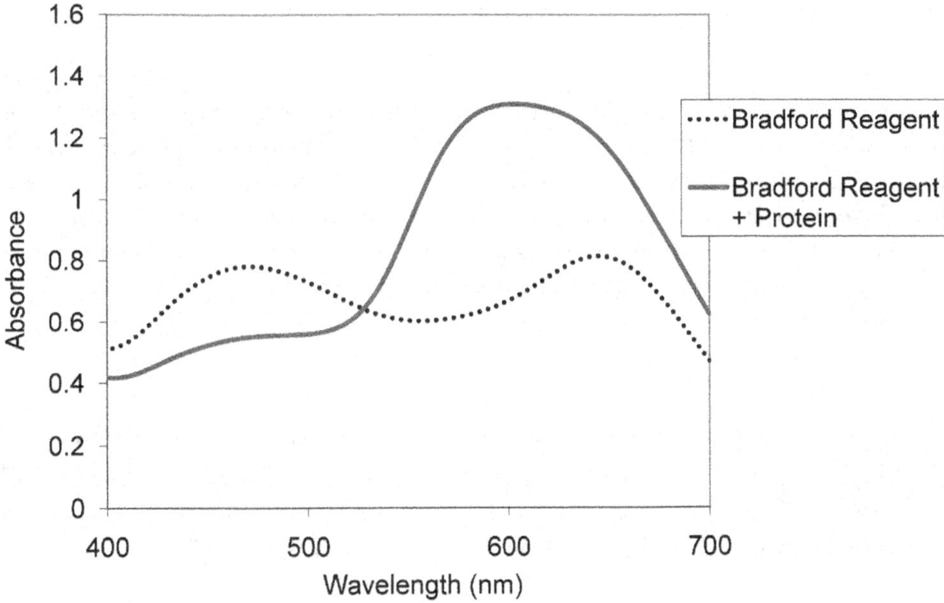


Figure 3.14 – Absorption spectra of Bradford Reagent and the reagent complexed with protein. Note a shift in wavelength as well as in absorbance occurs when protein is added to the reagent.

3) Lowry Assays

A substance containing two or more peptide bonds will complex with $Cu^{2+}$ ions, resulting in reduction of the copper ion. The $Cu^+$ ion and the oxidized amino acids (particularly aromatic amino acids) react with a second reagent (Folin-Ciocalteu Reagent, a phosphomolybdic-phosphotungstate complex) and reduce the phosphomolybdotungstate to heteropolymolybdenum blue. The amount of protein in the moderate to highly concentrated samples can be determined by measuring the absorbance of the sample at 550 nm. The complex also absorbs at 750 nm, and this absorbance is often used to determine lower protein concentrations (see Figure 3.15).

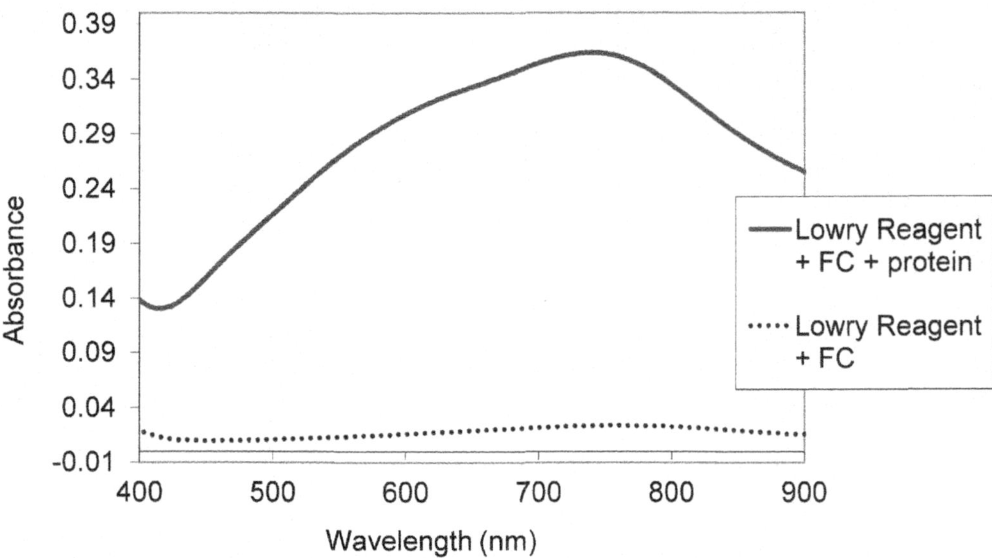


Figure 3.15 – Absorption spectra of the Lowry Reagent mixed with Folin-Ciocalteu and of the reagent complexed with protein. Note a shift in wavelength as well as in absorbance occurs when protein is added to the reagent.

The Lowry method is fairly sensitive; 0.1 – 1.0 mg/mL can be quantified using this assay. The method is slower than the Bradford assay or UV-Vis determination of concentration. The color develops after 20-30 minutes. Also, substances such as detergents, denaturants, buffers (such as Tris or high concentrations of acetate and phosphate), amino acids, ammonium sulfate, mercaptans, sugars and DNA interfere with the results of the assay. The color of the complex formed and the $\lambda_{max}$ may vary with the type of protein assayed.

4) Bicinchoninic Acid (BCA)

The BCA method is a variation of the Lowry assay. A protein sample is mixed with $Cu^{2+}$ ions and BCA (see Figure 3.16). The protein converts the copper from $Cu^{2+}$ to $Cu^+$. The $Cu^+$ is then chelated with the BCA (apple-green in color, $\lambda_{max}$ around 807 nm) to form a purple-colored copper-BCA complex, which absorbs light around 560 nm (see Figure 3.17).

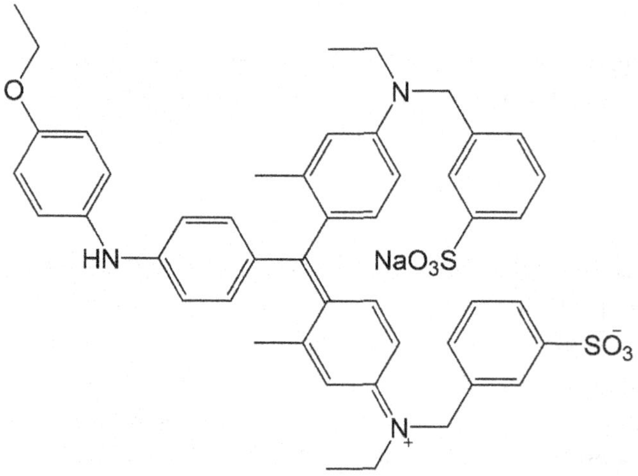

Figure 3.16 – Structure of bicinchonic acid.

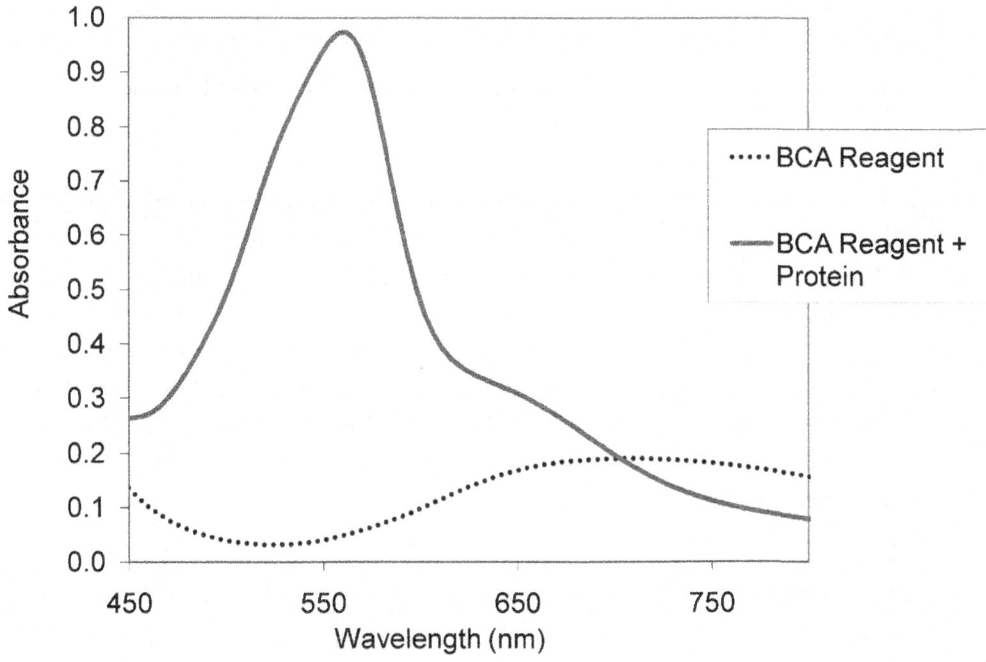

Figure 3.17 – Absorption spectra of the BCA Reagent as compared with the absorption spectra of the BCA Reagent missed with protein. Note a shift in wavelength as well as increase in absorbance occurs when the $Cu^+$ complexes with the BCA.

This method is also sensitive, detecting concentrations between 0.1 and 1.0 mg/mL of protein. It is compatible with detergents, but EDTA, dithiothretol, ammonium sulfate, and sugars interfere with this technique. The reagents are commercially available. Of the methods described in this chapter, the BCA method is very slow (depending on reaction conditions: the assay requires 40 minutes, 2 hours, or overnight to develop the color).

## 3.8 NUCLEIC ACID ASSAYS

The concentration of nucleic acids in solutions may be determined using various experimental techniques.

1) Spectrophotometric Assay

Nucleic acids strongly absorb ultraviolet light due to the presence of double bonds in the nitrogenous bases. Nucleic acids have a maximum absorbance at 260 nm, which is not affected by sugar phosphate configuration. The concentration of double-stranded DNA or single-stranded RNA or DNA can be determined from the $A_{260}$ value. Due to stacking of the bases and pairing of the DNA strands, the $A_{260}$ is lower for dsDNA than for ssDNA or for individual nucleotides; this is referred to as the hypochromic effect. An $A_{260}$ measurement of 1.0 (measured with a 1 cm pathlength) is equivalent to approximately 50 µg/mL of double-stranded DNA, 33 µg/mL of single-stranded DNA, or 40 µg/mL of single-stranded RNA (ref. Sambrook). For smaller molecules, such as oligonucleotides, the following equation can be used to calculate the molar extinction coefficient:

$$\varepsilon = (\text{number of A})(\varepsilon_A) + (\text{number of G})(\varepsilon_G) + (\text{number of C})(\varepsilon_C) + (\text{number of T})(\varepsilon_T)$$

$$\varepsilon = (\text{number of A})(15.3) + (\text{number of G})(11.9) + (\text{number of C})(7.4) + (\text{number of T})(9.3),$$

where $\varepsilon_A$, $\varepsilon_G$, $\varepsilon_C$, $\varepsilon_T$ are the molar extinction coefficients of the deoxyribonucleotide at pH 7.0 (ref. Sambrook).

Some researchers believe that this method can be used to determine the purity of the nucleic acid. Pure DNA samples should have a $A_{260}/A_{280}$ ratio of 1.8, and pure RNA samples have $A_{260}/A_{280}$ ratio of 2.0. Carbohydrates, peptides, phenols, or other aromatic compounds may absorb at 230 nm, and a ratio of $A_{260}/A_{230}$ should be approximately 2.2 for a nucleic acid sample free of these contaminants. If there is significant contamination with protein or phenol, the ratio will be less than 1.8 or 2.0, and a reliable prediction of the concentration of the nucleic acid will not be possible. The reliability of using this method to determine contamination has been questioned by some researchers because nucleic acids absorb strongly at 280 nm and 260 nm. Only a significant concentration of protein in the sample as a contaminant would change the absorbance $A_{260}/A_{280}$ ratio.

The advantages of this method are that it is nondestructive, rapid, and highly sensitive; the detection limit is approximately 2.5 μg/mL of nucleic acid. Disadvantages include interference from molecules such as proteins, EDTA, and phenol, and absorption coefficients that are affected by pH and ionic strength. The pH of the solution must be controlled and the ionic strength must be low.

2) Fluorometric Quantitation of DNA Using Hoechst 33258

The limitations of quantitation of double-stranded DNA by absorbance measurements, such as interferences by proteins and phenols and small quantity of isolated DNA, can be overcome by quantitation of DNA using fluorescence and Hoechst 33258. This fluorescent dye (shown in Figure 3.18) binds nonintercalatively to the minor groove of AT-rich segments of DNA, and the DNA-dye complex has a maximum excitation at 350 nm and emission at 450 nm. The sensitivity of the assay is approximately 1 ng/mL.

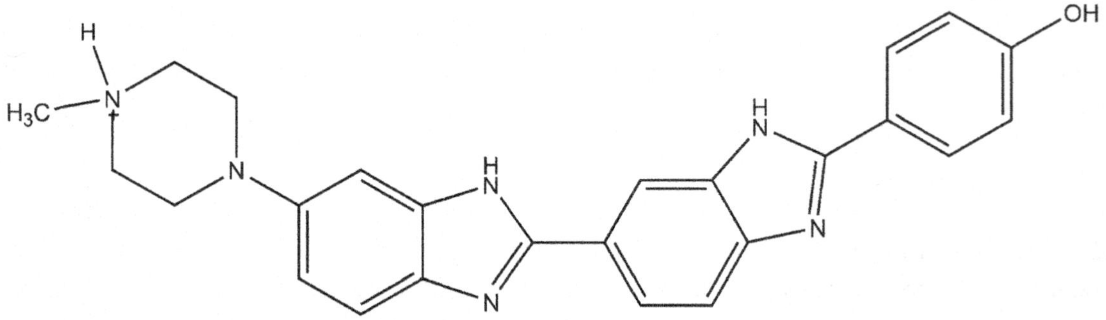

Figure 3.18 – Structure of Hoechst 33258 at neutral pH.

The binding of Hoechst 33258 to DNA is affected by the DNA sequence and conformation. It is important that the DNA sample used to create a standard curve for the assay has similar physical characteristics (AT content, conformation – supercoiled, relaxed, etc.) as the sample to be quantified. Calf thymus DNA is often used as a reference for plant and animal DNA samples because it contains approximately 58% AT.

The dye fluoresces only half as much when bound to single-stranded DNA or RNA as compared with double-stranded DNA. Also, short, single-stranded oligonucleotides will not form a complex with the dye or fluoresce in proportion to the concentration of the oligonucleotide.

Commonly used buffers to extract DNA have little to no effect on the assay or fluorescence properties. Low levels of detergent (<0.01% SDS) and salt up to 3 M NaCl have little effect on the assay. The dye has little affinity for proteins or RNA, and these contaminants would be less likely to interfere with the analysis. The DNA sample should be free of ethidium bromide because ethidium bromide quenches the fluorescence of Hoechst 33258.

3) Quantitation of DNA Using Ethidium Bromide

Ethidium bromide (shown in Figure 3.19) is a fluorescent dye that intercalates between the base pairs of DNA. The fluorescence of this dye increases 25-fold when intercalated into DNA. The amount of fluorescence is directly proportional to the amount of DNA or RNA in a sample. This method is frequently used if a rapid quantitation on small samples of DNA is required or if the DNA sample contains contaminants that absorb UV radiation.

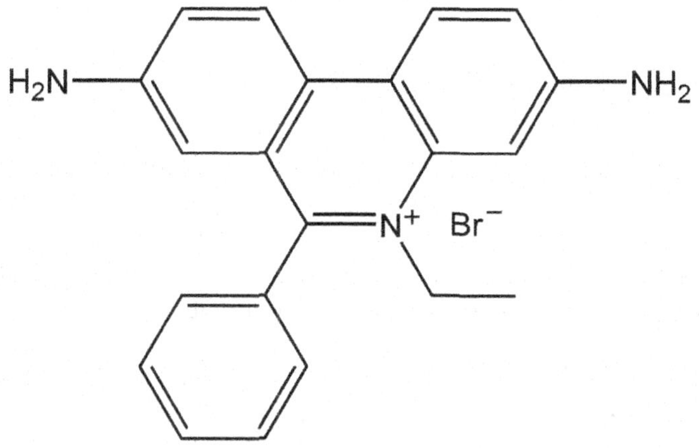

Figure 3.19 – Structure of ethidium bromide.

The use of ethidium bromide to detect DNA is fast and fairly sensitive, detecting 1-10 ng of nucleic acid. Disadvantages of this method include the fact that only a rough estimate of the DNA concentration can be made and that the method is useful for dilute samples of DNA. Above 15-20 ng of DNA, saturation is often reached, and the estimation is more difficult. Also, less ethidium bromide intercalates into a smaller DNA fragment than in larger DNA fragments. Therefore, the method may lead to an underestimation of the concentration of small DNA fragments.

## *Denaturation of DNA*

DNA exists in nature and solution as a double-stranded helix. When exposed to denaturing conditions (heat, alkali, or organic solvents), the hydrogen bonds between the base pairs are disrupted, and the DNA transitions from the double-stranded helical state to a random coil state. The double helix of DNA is stabilized by the hydrogen bonds between the base pairs and by hydrophobic and $\pi$-$\pi$ interactions due to stacking. As the base pairs are unstacked, the absorbance of UV light by the nucleic acid increases. This increase in absorbance is called the hyperchromic effect.

A thermal denaturation curve or melting curve for DNA can be obtained by plotting $A_{260}$ $_{(T)}/A_{280\ (25°C)}$ versus temperature (T) (ref. Boyer). The midpoint of the transition is called the melting temperature or transition temperature ($T_m$). The melting temperature is the temperature at which the DNA transitions from double-stranded helix to random coil and should not be confused with the transition from solid to liquid. At this point, 50% of the helical structure of DNA is lost, and the absorbance of UV light increases due to the hyperchromic effect. Each DNA has a characteristic $T_m$, and this value can be used for identification purposes.

## 3.9 QUANTITATIVE DETERMINATION OF CARBOHYDRATES IN SOLUTION

Carbohydrates do not contain intrinsic chromophores and, therefore, cannot be directly quantitated using UV-Vis or fluorescence spectroscopy. In order to quantitate carbohydrates using spectroscopy, one must chemically convert the carbohydrate to a derivative containing a chromophore or conduct a reaction between a carbohydrate and another substance in which a reagent changes color. Several spectroscopic methods of determining carbohydrates have been developed, and a few will be described below.

1) Dinitrosalicyclic Acid (DNSA) Reagent

The determination of carbohydrates using DNSA was originally developed in 1921 by Sumner and was later revised by Miller in 1959. Currently utilized protocols are revisions of Miller's published work. Although the majority of carbohydrate analysis used to date involves chromatographic separation followed by detection (see Chapter 7), several recently published journal articles utilize the DNSA reagent to quantitate glucose in samples.

The 3,5-dinitrosalicylic reagent reacts with the reducing ends of carbohydrates and converts from a yellow color (3,5-dinitrosalicylic acid) to a reddish-brown color (3-amino,5-nitrosalicylic acid) under alkaline conditions. The carbohydrate aldehyde or ketone functional group is oxidized in the reaction. Dissolved oxygen can interfere with the oxidation of a reducing sugar, and sulfite is added to the reagent to absorb dissolved oxygen and prevent air oxidation of the sugar. Phenol is often added to the reagent to intensify the color of the solution. It has been shown experimentally that the presence of phenol changes the slope of the calibration curve of absorbance versus reducing sugar concentration but does not affect the linearity of the curve. Potassium sodium tartrate (Rochelle salt) is added to stabilize the color.

The above-described reaction is the simplest explanation of what is occurring in the assay. It is suspected, however, that there are possible side reactions, and the actual reaction is more complicated. The type of side reaction depends on the exact nature of the reducing sugars. It has been reported that different reducing sugars yield varying amounts of color. Therefore, it is important to create a standard curve for different reducing sugars.

2) Anthrone Assay

The green- to blue-green-colored compound formed when 0.2% anthrone in concentrated sulfuric acid reacts with carbohydrates can be quantitated by recording the absorbance at 625 nm. It is believed that the carbohydrate is dehydrated in the strong acid to form a furfural, which condenses with the anthrone and forms the green compound. Studies have shown that the anthrone is somewhat unstable, and standard curves should be prepared each time the reagent is prepared to ensure that the assay obeys Beer's Law.

3) Resorcinol – Seliwanoff Reaction

Under acidic conditions, sugars dehydrate to form furfurals. The furfural compound then reacts with resorcinol, a colorless compound, to form a compound that is deep-red in color. Figure 3.20 shows the Seliwanoff reaction. The resulting red compound can be quantitated using UV-Vis spectroscopy or fluorescence.

The reaction can also be used to distinguish carbohydrates based on the type of carbonyl present in the compound. Ketose sugars react within 60 seconds and aldoses react in 2-5 minutes. Disaccharides may react under the same reaction conditions but at a much slower rate.

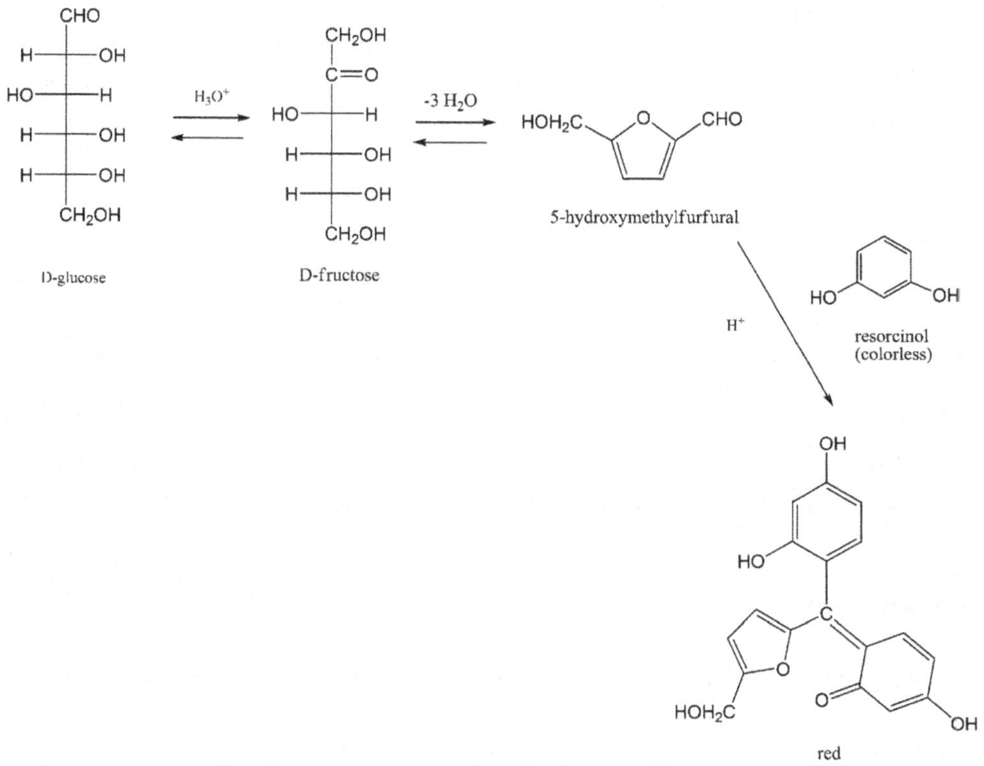

Figure 3.20 – Seliwanoff reaction.

# CONCLUSION

Spectroscopy is a very important tool utilized by biochemists and other scientists. This chapter provides an introductory look at the theory and practical uses of many forms of spectroscopic analysis. Additional information about each technique can be found in the many reference materials listed for this chapter.

# REFERENCES AND FURTHER READING

## Spectroscopy References

Pavia, D. L., Lampma, G. M., and G. S. Kriz. *Introduction to Spectroscopy. A Guide for Students of Organic Chemistry*. 3rd Edition. Orlando, FL: Harcourt College Publishing, 2001.

Sambrook, J., and D. W. Russell. *Molecular Cloning. A Laboratory Manual.* 3rd Edition. New York: Cold Springs Harbor Laboratory Press, 2001. Appendix 8.

## Denaturation of Proteins or DNA

Bastiras, S., and J. C. Wallace. Equilibrium denaturation of recombinant porcine growth hormone. *Bichemistry*, 31, 1992: 9304-9309.

Boyer, R. *Biochemistry Laboratory. Modern Theory and Techniques*. San Francisco: Pearson Benjamin Cummings, 2006.

Hargrove, M. S., and J. S. Olson. The Stability of Holomyoglobin Is Determined by Heme Affinity. *Biochemistry*, 35, 1996: 11310-11318.

Howard, K. P. Thermodynamics of DNA Duplex Formation. *J. Chem. Ed.* 77, 2000: 1469-1471.

Lau, F. W., and J. U. Bowie. A Method for Assessing the Stability of a Membrane Protein. *Biochemistry*, 36, 1997: 5884-5892.

Yu, X.-C., Shen, S., and H. W. Strobel. Denaturation of Cytochrome P450 2B1 by Guanidine Hydrochloride and Urea: Evidence for a Metastable Intermediate State of the Active Site. *Biochemistry*, 34, 1995: 5511-5517.

## Protein Quantification

Bollag, D. M., and S. J. Edelstein. Protein Methods. New York: Wiley-Liss, 1991.

Bradford, M. M. A Rapid and Sensitive Method for the Quantitation of Microgram Quantities of Protein Utilizing the Principle of Protein-Dye Binding. *Anal. Biochem.*, 72, 1976: 248-254.

Lowry, O. H., et al. Protein Measurements using Folin Phenol Reagent. *J. Biol. Chem.* 193, 1951: 265-275.

Pace, C.N., et al. How to Measure and Predict the Molar Absorption Coefficient of a Protein. *Protein Science*, 4, 1995: 2411-2423.

Smith, P.K., et al. Measurement of protein using bicinchoninic acid. *Anal. Biochem.*, 150, 1985: 76-85.

**Molar Absorption Coefficients of Proteins**

Edelhoch, H. Spectroscopic Determination of Tryptophan and Tyrosine in Proteins. *Biochemistry*, 6, 1976: 1948-1954.

Pace, C.N., et al. How to Measure and Predict the Molar Absorption Coefficient of a Protein. *Protein Sci.*, 4, 1995: 2411-2423.

Ramos, C. H. I. A Spectroscopic-based Laboratory Experiment for Protein Conformational Studies. *Biochem. Mol. Biol. Educ.*, 32, 2004: 31-34.

**Carbohydrate Quantification**

Miller, G. L. Use of Dinitrosalicylic acid Reagent for Determination of Reducing Sugar. *Anal. Chem.*, 31, 1959: 426-428.

Rogers, C. J., Chambers, C. W., and N. A. Clark. Rapid Spectrophotofluorometric Method for Determining Nanogram Quantities of Carbohydrates. *Anal. Chem.*, 28, 1966: 1851-1853.

Viles, F. J.,and L. Silverman. Determination of Starch and Cellulose with Anthrone. *Anal. Chem.*, 21, 1949: 950-953.

**IR Spectrometry**

Petibois, C., et al. Plasma Protein Contents Determined by Fourier-Transform Infrared Spectrometry. *Clin. Chem.*, 45, 1999: 1530 - 1535.

Petibois, C., et al. Determination of Glucose in Dried Serum Samples by Fourier-Transform Infrared Spectroscopy. *Clin. Chem.*, 45, 1999: 1530 - 1535.

Sockalingum, G. D., et al. FT-IR Spectroscopy as an Emerging Method for Rapid Characterization of Microorganisms. *Cell Mol. Biol.*, 44, 1998: 261-269.

**NMR Spectrometry**

Den Hollander, J. A., et al. [13]C Nuclear Magnetic Resonance Studies of Anaerobic Glycolysis in Suspensions of Yeast Cells. *Proc. Nat. Acad. Sci.*, 76, 1979: 6096-6100.

Duarte, I., et al. High-Resolution Nuclear Magnetic Resonance Spectroscopy and Multivariate Analysis for the Characterization of Beer. *J. Agric. and Food Chem.*, 50, 2002: 2475-2481.

Mega, T. L., Carlson, C. B., and D. A. Cleary. Following Glycolysis using [13]CNMR. *J. Chem. Educ.*, 74, 1997: 1474-1476.

Wüthrich, K. The Way to NMR Structures of Proteins. *Nature Struc. Biol.*, 2001: 923-925

**Mass Spectrometry**

Aebersold, R., and D. R. Goodlett. Mass Spectrometry in Proteomics. *Chem. Rev.,* 101, 2001:
    269-296.

Counterman, A. E., Thompson, M.S., and D. E. Clemmer. Identifying a Protein by MALDI-
    TOF Mass Spectrometry. *J. Chem. Educ.,* 80, 2003: 177-180.

Muddiman, D.C., et al. Matrix-Assisted Laser Desorption/Ionization Mass Spectrometry. *J.
    Chem. Educ.,* 74, 1997: 1288-1292.

**Online References**

"Protein, gene searches." National Center for Biotechnology Information. *Pubmed.*
    <http://www.ncbi.nlm.nih.gov/>.


*Spectroscopy now.* Free access site with journal articles about spectroscopic techniques.
    <http://www.spectroscopynow.com/>.

# Experiment 5. Spectrophotometric Studies of Proteins

## Purpose of Experiment

In this experiment, you will prepare a standard curve and determine the concentration of an unknown protein mixture using several different techniques. You will determine the most accurate way to measure protein concentrations. You will also calculate the molar absorptivity coefficients of the proteins using two pieces of information: the amino acid composition and the absorbance data.

## Prelaboratory Assignment

Complete the following exercises and/or answer the following questions:

1) Prior to arriving in lab, each student should plan the experiments to be conducted. The procedures below provide guidelines for the preparation of samples, but the student should calculate the volume or amounts of each substance to use for each experiment.

2) Based on the information presented in the introduction and/or other knowledge, predict which method of protein concentration determination is most accurate.

## Materials

Stock solution of bovine serum albumin (BSA) or other standard protein (1 mg/mL or 2 mg/mL concentration).

Bradford working buffer (Dissolve 10 mg Coomassie Brilliant Blue G-250 in 5 ml 95% ethanol, add 10 ml 85% (w/v) phosphoric acid. Dilute to 100 mL when the dye has completely dissolved and filter through Whatman #1 paper just before use.)

Lowry Reagent 1: Mix one volume of Reagent B (0.5% copper sulfate pentahydrate, 1% sodium or potassium tartrate) with 50 volumes of Reagent A (2% sodium carbonate, 0.4% NaOH). Reagent A and B can be stored indefinitely at room tempearature.

Lowry Reagent 2: Dilute commercial Folin-Ciocalteu phenol reagent with an equal volume of water. This solution is stable for a few days or weeks.

BCA Reagent A (1% bicinchoninic acid disodium salt ($Na_2BCA$, 2% sodium carbonate, 0.16% sodium tartrate, 0.4% sodium hydroxide, and 0.95% sodium bicarbonate; adjust pH of solution to 11.25 with NaOH or $NaHCO_3$)

BCA Reagent B (4% copper sulfate pentahydrate). Reagents A and B are commercially available and are stable for at least 1 year at room temperature.

BCA Working Reagent: Mix 50 volumes of BCA Reagent A with 1 volume of BCA Reagent B. This solution is stable for one week.

**NOTE** – with spectroscopic measurements, inspect cuvettes for scratches or damage, ensure that the proper volume is used and that the sample is free of suspended debris or air bubbles that could cause scattering of light and provide inaccurate results.

## Spectrometric Assay

1) Turn on the spectrophotometer and allow the instrument to warm up for the time specified by the manufacturer.

2) Following the instructions for the spectrophotometer, obtain a background scan from 220–300 nm using experimental buffer (in which protein is dissolved) or solvent (typically water).

3) Replace the buffer in the sample cuvette with the protein sample. Obtain a scan from 220–300 nm. Determine the absorbances at 260 nm, 280 nm, and 230 nm.

4) If the absorbance at 280 nm is greater than 1 or 2, dilute the protein sample quantitatively and repeat Step 3 until the absorbance is less than 1 or 2.

## Bradford Assay

1) Create a dilution series of calibration samples containing aliquots of BSA or other standard protein at increasing amounts (0 to 20 µg) (maximum sample volume of 100 µL). Bring the volume of each sample to 100 µL with experimental buffer. Samples should be prepared in duplicate or triplicate to ensure accuracy.

2) If an approximate concentration of the unknown protein is known, prepare a sample containing around 10-15 µg of protein in a 100 µL sample. If the approximate concentration is unknown, prepare samples containing 25 µL and 50 µL of the provided solution of protein diluted with enough buffer to create 100 µL of solution.

3) Prepare a sample for blanking the instrument containing only 100 µL of water. Depending on the type of spectrophotometer used, duplicates of this sample may be needed.

4) Add 2.0 mL of Bradford reagent solution to each sample. Mix thoroughly. Allow to sit for 5 minutes at room temperature. Measure the absorbances of the solutions after 5 but no longer than 60 minutes after mixing. Precipitation of a protein-Coomassie complex frequently begins after 15 minutes. This can interfere with the accuracy of the method.

5) Using plastic or glass cuvettes, blank the spectrophotometer using the sample containing only water and Bradford reagent at 595 nm. After each use, rinse the cuvettes with ethanol and then water to remove residual dye.

6) Record absorbances for all other samples at 595 nm.

7) Prepare a standard curve of absorbance at 595 nm versus µg of BSA (or standard protein used).

8) Use the standard curve to approximate concentration of protein in unknown sample(s).

## Lowry Assay

1) Create a dilution series of calibration samples containing aliquots of BSA or other standard protein at increasing amounts (0 to 80 μg) (maximum sample volume of 500 μL). Bring the volume of each sample to 500 μL with experimental buffer. Samples should be prepared in duplicate or triplicate to ensure accuracy.

2) If an approximate concentration of the unknown protein is known, prepare a sample containing around 40-60 μg of protein in a 500 μL sample. If the approximate concentration is unknown, prepare samples containing 100 μL and 200 μL of the provided solution of protein diluted with enough buffer to create 500 μL of solution.

3) Prepare a sample for blanking the instrument containing only 500 μL of water. Depending on the type of spectrophotometer used, duplicates of this sample may be needed.

4) Add 2.5 mL of Lowry reagent 1 to each sample. Mix thoroughly. Allow to stand for 5-10 minutes at room temperature.

5) Add 250 μL of Lowry reagent 2 and mix thoroughly. Allow the samples to incubate for 20-30 minutes at room temperature.

6) Blank the spectrophotometer using sample containing Lowry reagents and water. Record absorbances for all samples at 750 nm.

7) Prepare a standard curve of absorbance at 750 nm versus μg of BSA (or standard protein used).

8) Use the standard curve to approximate concentration of protein in unknown sample(s).

## BCA Assay

1) Create a dilution series of calibration samples containing aliquots of BSA or other standard protein at increasing amounts (0–80 μg) (maximum sample volume of 100 μL). Bring the volume of each sample to 100 μL with experimental buffer. Samples should be prepared in duplicate or triplicate to ensure accuracy.

2) If an approximate concentration of the unknown protein is known, prepare a sample containing around 40–60 μg of protein in a 100 μL sample. If the approximate concentration is unknown, prepare samples containing 50 μL and 100 μL of the provided solution of protein diluted with enough buffer to create 100 μL of solution.

3) Prepare a sample for blanking the instrument containing only 100 μL of water. Depending on the type of spectrophotometer used, duplicates of this sample may be needed.

4) Mix each sample with 2 mL of BCA working reagent. Mix thoroughly. Allow to stand for 2 hours at room temperature or incubate at 37°C for 30 minutes or at 4°C overnight. Allow to cool or warm to room temperature (if necessary).

5) Blank the spectrophotometer using the sample containing BCA reagent and water. Record absorbances for all samples at 562 nm.

6) Prepare a standard curve of absorbance at 562 nm versus µg of BSA (or standard protein used). (See Figure 3.2.)

7) Use the standard curve to approximate concentration of protein in unknown sample(s).

## Determination of Molar Extinction Coefficient of Protein

1) Obtain a solid sample or stock solution of a protein from the instructor.

2) Create a Beer's law plot for the protein by measuring the absorbance at 280 nm at various concentrations of the protein (varying between 0–2 mg/mL).

3) Using a literature or BLAST search, determine the sequence and molecular weight of the protein.

4) Using the slope of the line and the molecular weight of the protein, determine the molar extinction coefficient of the protein.

5) Calculate the molar extinction coefficient using the equation:

$$\varepsilon_{\text{protein at 280 nm}} = n_{\text{Trp}}\varepsilon_{\text{Trp at 280 nm}} + n_{\text{Tyr}}\varepsilon_{\text{Tyr at 280 nm}} + n_{\text{Cystine}}\varepsilon_{\text{Cystine at 280 nm}}$$

(See Table 3.2.)

6) Compare the extinction coefficients.

## Optional Additional Explorations of Protein Quantitification

Design experiments to investigate the effects of interferences, such as buffer components, other macromolecules, etc., on the protein assays described above. Alternatively, explore the effect of different protein structures (different sequences) on the sensitivity and accuracy of each assay. Prior to any investigations, seek approval for the procedures from your instructor.

## Questions

1) What chemical features of proteins are measured by each assay? How specific is each assay? Which assay would be more useful for approximating total protein in a sample? How do your answers relate to your prelaboratory predictions?

2) What substances might interfere with these assays? Which assay would be better to determine concentration of protein in the presence of other biological macromolecules?

3) Comment on the accuracy and precision of each of the assays.

4) Does the structure of the protein affect its spectrophotometric properties?

## References for the Experiments

Bollag, D. M.,  and S. J. Edelstein. *Protein Methods.* New York: Wiley-Liss, 1991.

Bradford, M. M. A Rapid and Sensitive Method for the Quantitation of Microgram Quantities of Protein Utilizing the Principle of Protein-Dye Binding. *Anal. Biochem.,* 72, 1976: 248-254.

Edelhoch, H. Spectroscopic Determination of Tryptophan and Tyrosine in Proteins. *Biochemistry,* 6, 1967: 1948-1954.

Lowry, O. H., et al. Protein Measurements using Folin Phenol Reagent. *J. Biol. Chem.,* 193, 1951: 265-275.

Pace, C. N., et al. How to Measure and Predict the Molar Absorption Coefficient of a Protein. *Protein Science,* 4, 1995: 2411-2423.

Ramos, C. H. I. A Spectroscopic-based Laboratory Experiment for Protein Conformational Studies. *Biochem. Mol. Biol. Educ.,* 32, 2004: 31-34.

Smith, P. K., et al. Measurement of protein using bicinchoninic acid. *Anal. Biochem.,* 150, 1985: 76-85.

# Experiment 6. Spectrophotometric Studies of Nucleic Acids in Solution

## Purpose of Experiment

In this experiment, you will determine the concentration of an unknown nucleic acid using several different techniques. You will also determine the melting temperature of DNA under various conditions.

## Prelaboratory Assignment

Complete the following exercises and/or answer the following questions:

1) Prior to arriving in lab, each student should plan the experiments to be conducted. The procedures below provide guidelines for the preparation of samples, but the student should calculate the volume or amounts of each substance to use for each experiment.

2) Based on the information presented in the introduction and/or other knowledge, predict which method of DNA concentration determination is most accurate.

## Materials

Calf thymus DNA standard solution (1 mg/mL) or other DNA standard (such as herring sperm) in TE buffer

Unknown DNA sample

TE buffer, pH 7.6 (10 mM Tris-HCl, 1 mM EDTA (pH 8))

Hoechst 33258 stock dye solution (1 mg/mL): Dissolve 10 mg of Hoechst 33258 in enough filtered water to make 10 mL of solution. Store in an amber bottle at 4°C for up to 6 months.

10X TNE buffer stock solution: Dissolve 12.11 g Tris base, 3.72 g EDTA, disodium salt, dehydrate, 116.89 g Sodium chloride in approximately 800 mL of distilled water. Adjust pH to 7.4 with concentrated HCl, and then add enough distilled water to create 1000 mL of solution. Filter the solution using a 0.45 µm filter before use. The solution can be stored at 4°C for up to 3 months.

1X TNE: Dilute one portion 10X TNE with 9 parts of distilled water

2X Dye Solution (200 ng/mL) for 10-1000 ng/mL final DNA concentration:
Dilute 20 µL Hoechst 33258 stock solution (1 mg/mL) with 100 mL 1X TNE. Keep assay solution at room temperature. Prepare fresh daily. Do not filter once dye has been added.

Quartz cuvettes

Spectrophotometer

Water bath

Cuvettes

Thermometer

Tris buffer I (0.01 M Tris-HCl and 0.05 M NaCl, pH 7.5)

Phosphate buffer (0.01 M $Na_2HPO_4$ / $Na_3PO_4$, pH 10)

Methanol
Tris buffer II (0.01 M Tris-HCl and 1.0 M NaCl, pH 7.5)

**NOTE –** with spectroscopic measurements, inspect cuvettes for scratches or damage and ensure that the proper volume is used and that the sample is free of suspended debris or air bubbles that could cause scattering of light and provide inaccurate results.

## Estimation and Characterization of DNA - Spectrometric Assay

1) Dilute a 0.5 mL aliquot of standard DNA solution with 4.5 mL of TE buffer.

2) Following the instructions for the spectrophotometer, obtain a background scan from 220–300 nm using Tris buffer.

3) Replace the buffer in the sample cuvette with the diluted DNA sample from Step 1. Obtain a scan from 220–300 nm. Determine the absorbance at 260 nm, 280 nm, and 230 nm.

4) Repeat with the unknown DNA sample.

## Fluorescent Quantification Using Hoechst 33258

Safety Precaution: Hoechst 33258 is a potential carcinogen and possible mutagen. Users should wear gloves and a mask and work under a fume hood when handling the dye.

## Protocol

1) Create a dilution series of calibration samples with the standard DNA samples in TE buffer with varying concentrations between 0 and 2 µg/mL. Suggested concentrations in ng/mL are 0, 25, 50, 100, 250, 500, 1000, and 2000. The size of the sample will depend on the size of the cuvettes (see Step 2).

2) Mix equal volumes of the DNA solution in Step 1 with an equal volume of the 2X dye solution. The final volume of the mixture should be adequate for the cuvettes for the fluorometer. For example, if the cuvettes hold 200 µL, a 100 µL sample of DNA would be mixed with 100 µL of the 2X dye.

3) Turn on the fluorometer and allow the instrument to warm up for the time recommended by the manufacturer.

4) Set up the parameters and calibrate the fluorometer (refer to instrumental operation manual). Recall that Hoechst 33258 is excited at 350 nm and emits at 450 nm.

5) Measure the fluorescence of each sample.

6) Create a standard curve of DNA fluorescence versus concentration of DNA.

7) Mix equal volumes of the unknown DNA solution with an equal volume of the 2X dye solution.

8) Use the fluorometer to measure fluorescence of the unknown DNA sample. Estimate the concentration of DNA in the unknown using interpolation on the standard curve. If the reading for the unknown DNA falls outside the linear range of the standard curve, a sample with greater concentration of the unknown or a dilution of the original unknown should be prepared.

## Determination of the Melting Temperature of DNA Sample

Safety Precaution: Care should be taken when handling hot glassware.

1) Blank a spectrophotometer at 260 nm with the Tris buffer I.

2) Dilute 100 µL of the standard DNA solution with 3 ml of Tris buffer I. Measure the $A_{260}$ and $A_{280}$ of the sample at room temperature.

3) Measure the temperature of the sample in the cuvette.

4) Place the cuvette in a water bath on a hot plate, leaving the thermometer in the cuvette. Slowly heat the sample. Periodically (approximately every 5°C), record the temperature and quickly (but carefully) remove the cuvette, wipe the sides, and measure the absorbance at 260 nm.

5) Repeat the heating/recording $A_{260}$ until the sample reaches around 90°C.

6) Plot $A_{260\ (T)}\ /\ A_{280(25°C)}$ versus temperature to determine $T_M$.

7) Repeat Steps 1-6, except use 50% methanol in water as the solvent for the blanking and dilution. This sample can only be heated until around 80°C, as the solution will begin to boil.

8) Repeat Steps 1-6, except use Tris buffer II as the solvent.

9) Repeat Steps 1-6, except use phosphate buffer at pH 10 as the solvent.

10) An alternative procedure can be used if the spectrophotometer has a heating attachment. Simply, monitor $A_{260}$ as the temperature is increased from 25–90°C. It may also be possible to wrap the cuvette holder in heating tape and increase the temperature using a power supply.

## Optional Additional Explorations of DNA Quantitification

Design experiments to investigate the effects of interferences, such as buffer components, other macromolecules, etc., on the DNA assays described above. Alternatively, explore the effect of different nucleic acid structures (different sequences) on the sensitivity and accuracy of each assay. Prior to any investigations, seek approval for the procedures from your instructor.

## Questions

1)  What chemical features of DNA are measured by each assay? How specific is each assay? Which assay would be more useful for approximating total DNA in a sample? How does your answer agree with your prelaboratory prediction?

2)  What substances might interfere with these assays? Which assay would be better to determine concentration of DNA in the presence of other biological macromolecules?

3)  Comment on the accuracy and precision of each of the assays.

4)  Calculate $A_{260}/A_{280}$ and $A_{260}/A_{230}$ ratios. Comment on the purity of your nucleic acid sample.

5)  Does the structure of the DNA affect the melting temperature?

6)  What is the melting temperature of your DNA sample under each of the conditions? Based on your understanding of the structure of DNA, do your experimental findings make sense?


## References for the Experiments

Boyer, R. F. *Modern Experimental Biochemistry.* 3$^{rd}$ edition. Redwood City: Benjamin/Cummings Publishing Company, 2000.

Howard, K. P. Thermodynamics of DNA Duplex Formation. *J. Chem. Ed.,* 77, 2000: 1469-471.

Sambrook, J., and D. W. Russell. *Molecular Cloning. A Laboratory Manual.* 3$^{rd}$ Edition. New York: Cold Springs Harbor Laboratory Press, 2001. Appendix 8.

"Turner Biosystems A Modulus™ Method for DNA Quantitation using Hoecsht Dye 33258, an application file." 10 Jan. 2006. http://www.turnerbiosystems.com/doc/appnotes/S_0118.php.

# Experiment 7. Determination of Concentration of Carbohydrates in Solution

## Purpose of Experiment

In this experiment, you will prepare a standard curve and determine the concentration of an unknown carbohydrate mixture using several different techniques. You will evaluate the most accurate way to measure concentrations of sugars in solution.

## Prelaboratory Assignment

Complete the following exercises and/or answer the following questions:

1) Prior to arriving in lab, each student should plan the experiments to be conducted. The procedures below provide guidelines for the preparation of samples, but the student should calculate the volume or amounts of each substance to use for each experiment.

2) Based on the information presented in the introduction and/or other knowledge, predict which method of carbohydrate concentration determination is most accurate.

## Materials

Stock solution of carbohydrate/sugars
Dinitrosalicylic acid (DNSA) reagent (10 g 3,5-dinitrosalicylic acid (DNSA), 2 g phenol
      (optional), 0.5 g sodium sulfite, 10 g sodium hydroxide, water to 1 L)
Potassium sodium tartrate or potassium tartrate solution, 40%
Anthrone reagent (0.2% anthrone in concentrated sulfuric acid); prepare fresh
Resorcinol (50 mg resorcinol dissolved in 20 mL of concentrated HCl)

**NOTE** – with spectroscopic measurements, inspect cuvettes for scratches or damage and ensure that the proper volume is used and that the sample is free of suspended debris or air bubbles that could cause scattering of light and provide inaccurate results. The solutions used involve strong acids. Make sure that the cuvettes are rinsed well with water to remove residual acid.

Safety Precaution: Concentrated acids are used in this laboratory exercise. Avoid contact with skin and avoid breathing vapors. Wear gloves and dispose of solutions as instructed. Do not place concentrated acids, especially sulfuric acid, in plastic containers.

DNSA Assay

1) Create samples containing aliquots of carbohydrates at increasing concentrations (0–800 μg) in 3 mL sample. Samples should be prepared in duplicate or triplicate to ensure accuracy.

2) Prepare a sample of the unknown carbohydrate following the same procedure as in Step 1.

3)  Add an equal volume of DNSA reagent to each sample (those from Steps 1 and 2) and heat in a boiling water bath for 10 minutes.

4)  Add 1 mL of 40% potassium sodium tartrate.

5)  Cool the sample to room temperature.

6)  Blank the spectrophotometer. Record absorbances for all samples at 575 nm.

7)  Prepare a standard curve of absorbance at 575 nm versus concentration of carbohydrate (in μg).

8)  Use the standard curve to approximate concentration of carbohydrate in samples.

Anthrone Assay

1)  Create samples containing aliquots of carbohydrates at increasing concentrations (0–80 μg). (Ensure that each has the same volume.) Samples should be prepared in duplicate or triplicate to ensure accuracy.

2)  Prepare a sample of the unknown carbohydrate following the same procedure as in Step 1.

3)  Add two volumes of anthrone reagent to each sample and heat in a boiling water bath for 20 minutes.

4)  Cool the sample to room temperature.

5)  Blank the spectrophotometer. Record absorbances for all samples at 625 nm.

6)  Prepare a standard curve of absorbance at 625 nm versus concentration of carbohydrate (in μg).

7)  Use the standard curve to approximate concentration of carbohydrate in samples.

Resorcinol Assay

1)  Create samples containing aliquots of carbohydrate at increasing concentrations (0–80 μg) (maximum volume of 100 μL). Bring the volume of each sample to 100 μL with water. Samples should be prepared in duplicate or triplicate to ensure accuracy.

2)  Prepare a sample of the unknown carbohydrate following the same procedure as in Step 1.

3)  Add 0.5 mL of concentrated HCl and 50 μL of resorcinol solution to each sample. Mix thoroughly.

4)  Heat samples in a boiling water bath for 30 minutes.

5)  Cool samples to room temperature.

6) Adjust the pH to 8.5-9.5 using NaOH, and dilute the samples with water to 10 mL.

7) Blank the spectrofluorometer using sample with zero carbohydrate. Record fluorescence for all samples at 508 nm using excitation at 488 nm.

8) Prepare a standard curve of fluorescence intensity versus concentration of carbohydrate (in μg).

9) Use the standard curve to approximate concentration of sugar in samples.

## Optional Additional Explorations of Carbohydrate Quantitification

Design experiments to investigate the effects of interferences, such as buffer components, other macromolecules, etc., on the carbohydrate assays described above. Prior to any investigations, seek approval for the procedures from your instructor.

## Questions

1) What chemical features of carbohydrates are measured by each assay? How specific is each assay? Which assay would be more useful for approximating total carbohydrate in a sample? How do your answers relate to your prelaboratory predictions?

2) Could all carbohydrate samples be analyzed using these techniques?

3) What substances might interfere with these assays? Which assay would be better to determine concentration of carbohydrate in the presence of other biological macromolecules?

4) Comment on the accuracy and precision of each of the assays.

## References for the Experiment

Miller, G. L. Use of Dinitrosalicylic acid Reagent for Determination of Reducing Sugar. *Anal. Chem.*, 31, 1959: 426-428.

Rogers, C. J., Chambers, C.W., and N. A. Clark. Rapid Spectrophotofluorometric Method for Determining Nanogram Quantities of Carbohydrates. *Anal. Chem.*, 38, 1966: 1851-1853.

Viles, F. J., and L. Silverman. Determination of Starch and Cellulose with Anthrone. *Anal. Chem.*, 21, 1949: 950-953.

# Experiment 8. Identification of Carbohydrates

## Purpose of the Experiment

The purpose of the experiment is to identify the carbohydrate present in an unknown solution and investigate the chemical properties of carbohydrates. Numerous experimental tests that are used to identify various chemical features in carbohydrates are provided in the introduction and experimental procedure section of this laboratory handout. You will need to decide which tests you will conduct in the laboratory to assist you in identifying your unknown. Unless specified by your instructor, you do not need to use all of these tests, and you can use other tests as long as your instructor approves them prior to conducting the experiment. A flow chart of the protocol for identification of the unknown may assist you in determining the identity of your unknown sample.

## Prelaboratory Assignment

1) Draw the structures of the known sugars for this experiment. Structures should be drawn in the straight chain and in the cyclic forms.

2) Identify the chemical features of each sugar (i.e., ketose or aldose, number of carbons, etc.) What are anomers? Which sugars can have anomers? How does that relate to reducing sugars?

3) Identify which compounds will react positively with each test.

4) Develop a flow chart for the experiment.

## Background

Carbohydrates are very abundant biological molecules. The basic units of carbohydrates are monosaccharides. Monosaccharides differ in their structures, and these differences can be used to determine certain characteristics of the sugar. Various laboratory tests exist to identify carbohydrates and are based on these structural differences.

1) Barfoed's Test

Barfoed's test is used to distinguish between reducing monosaccharides and reducing disaccharides. The reducing sugar is oxidized to the carboxylic acid and $Cu^{2+}$ is reduced to $Cu^+$. The reduction of the monosaccharide occurs more rapidly than for disaccharides. The appearance of the red precipitate (copper (I) oxide) confirms the presence of a reducing sugar. A reducing monosaccharide can be distinguished from a reducing disaccharide based on the rate of appearance of the red precipitate. Ketose sugars first tautomerize to the aldose sugar and then react with the $Cu^{2+}$ ion. Monosacharides will react within 2-3 minutes, whereas disaccharides react in greater than 10 minutes. A nonreducing sugar will slowly hydrolyze and then react.

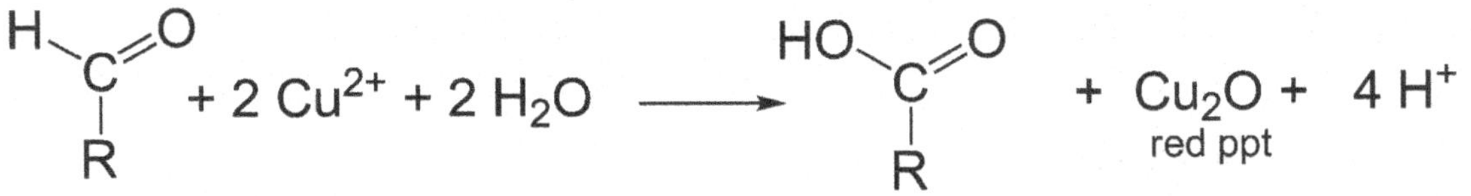

$$H-\overset{\displaystyle O}{\underset{\displaystyle R}{C}} + 2\ Cu^{2+} + 2\ H_2O \longrightarrow HO-\overset{\displaystyle O}{\underset{\displaystyle R}{C}} + Cu_2O + 4\ H^+$$

red ppt

2) Benedict's Test

The Benedict's test also differentiates between reducing and nonreducing sugars but cannot distinguish monosaccharides from disaccharides. The difference in the reactions is that the Barfoed test occurs under acidic conditions, whereas the Benedict's test occurs under basic conditions. Citrate is present in the reaction mixture to prevent the precipitation of $Cu(OH)_2$. Reducing sugars will react to form a red, brown, or yellow precipitate.

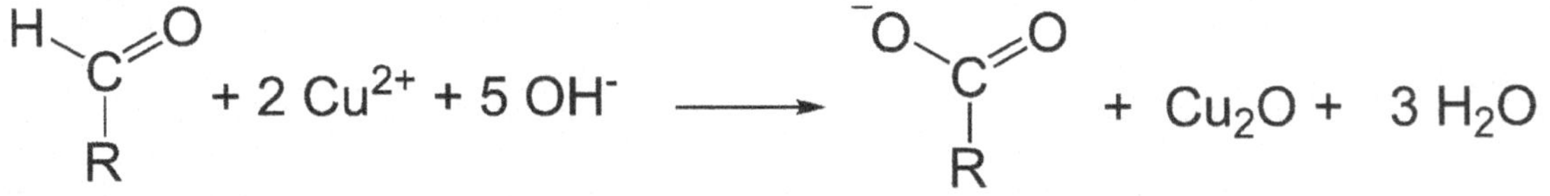

$$H-\overset{\displaystyle O}{\underset{\displaystyle R}{C}} + 2\ Cu^{2+} + 5\ OH^- \longrightarrow {}^-O-\overset{\displaystyle O}{\underset{\displaystyle R}{C}} + Cu_2O + 3\ H_2O$$

3) Bial Test

The Bial test is used to differentiate between pentose and hexose sugars. Under acidic conditions, pentose sugars are rapidly dehydrated to form furfural. The furfural then reacts with orcinol and ferric-chloride to form a blue-green solution.

Six-carbon sugars are also dehydrated under acidic conditions but form 5-hydroxymethylfurfural instead of furfural. 5-hydroxymethylfurfural reacts with orcinol and $FeCl_3$, but the reaction occurs at a much slower rate and forms reddish-brown complexes. Disaccharides can also react with Bial reagent. Under acidic conditions, the glycosidic bond is cleaved, and the resulting monosaccharides can react to form furfural or 5-hydroxymethylfurfural.

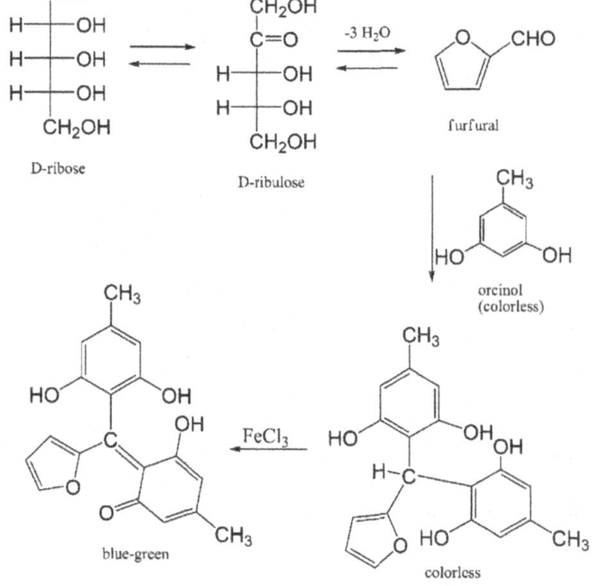

4)  Elson-Morgan Determination of Hexosamines

Analysis of 2-deoxy-2-amino sugars involves condensation of the sugar with 2,4-penta-dione in a basic solution. The product (a pyrrole derivative) is then reacted with $p$-dimethylaminobenzaldehyde in acidic solution to form a chromogen that has a maximum absorption at 530 nm. 6-deoxy-6-aminosugars and 3-deoxy-3-substituted amino acids also give rise to positive results but form compounds that have maximum absorptions at other wavelengths.

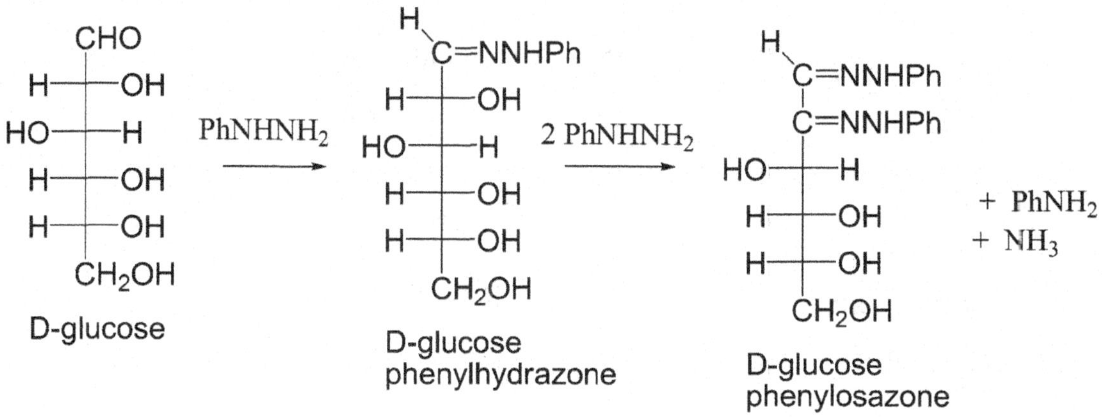

$$
\begin{array}{ccccc}
\text{CHO} & & \text{H}\!-\!\text{C}=\text{NNHPh} & & \text{H}\!-\!\text{C}=\text{NNHPh} \\
\text{H}\!-\!\text{OH} & & \text{H}\!-\!\text{OH} & & \text{C}=\text{NNHPh} \\
\text{HO}\!-\!\text{H} & \xrightarrow{\text{PhNHNH}_2} & \text{HO}\!-\!\text{H} & \xrightarrow{\text{2 PhNHNH}_2} & \text{HO}\!-\!\text{H} \\
\text{H}\!-\!\text{OH} & & \text{H}\!-\!\text{OH} & & \text{H}\!-\!\text{OH} \\
\text{H}\!-\!\text{OH} & & \text{H}\!-\!\text{OH} & & \text{H}\!-\!\text{OH} \\
\text{CH}_2\text{OH} & & \text{CH}_2\text{OH} & & \text{CH}_2\text{OH} \\
\text{D-glucose} & & \begin{array}{c}\text{D-glucose}\\\text{phenylhydrazone}\end{array} & & \begin{array}{c}\text{D-glucose}\\\text{phenylosazone}\end{array}
\end{array}
\qquad + \text{PhNH}_2 + \text{NH}_3
$$

5)  Osazone Formation

Osazones (also referred to as phenylosazones) are formed in the reaction of a sugar with phenylhydrazine.

Osazones are solid derivatives of the sugars and could be used to determine the identity of the sugar based on the melting point of the derivative. However, the osazones melt over a very broad range and identical phenylosazones are formed from different sugars. D-glucose, D-mannose, and D-fructose all give rise to the phenylosazone shown above. Some sugars can be distinguished by examining the rate at which the osazone forms and the appearance of the crystals. For example, fructose reacts faster than glucose, and glucose produces coarse crystals, wherease arabinose produces a product with fine crystals.

6)   Red Tetrazolium Test

The red tetrazolium (2,3,5-triphenyl-2H-tetrazolium chloride) test is a simple test for the presence of a reducing sugar. In the presence of a reducing sugar, the nearly colorless reagent is reduced to form a diformazan compound, an intensely colored pigment. The test can be used to distinguish ketose from aldose sugars based on the reaction times.

$$\text{Red tetrazolium} + 2H \longrightarrow \text{RT-Diformazan} + 2\,HCl$$

Red tetrazolium                                RT-Diformazan

7)   Seliwanoff Test

The Seliwanoff test is used to distinguish between aldohexoses and ketohexoses. See Section 3.9 of the text for details on this test.


## Materials

Known carbohydrate samples (suggestions: arabinose, fructose, glucose, lactose,
        maltose,sucrose, N-acetylglucosamine, galactose, ribose)
Unknown carbohydrate sample provided by instructor
Barfoed's Reagent (6.7 g of copper (II) acetate and 0.9 mL of glacial acetic acid in 100 ml
        of water)
Benedict's Reagent (100 g $Na_2CO_3$, 175 g sodium citrate, and 17.3 g of $CuSO_4 \cdot 5H_2O$
        per liter)
Bial's Reagent (300 mg of 3,5-dihydroxytoluene dissolved in 100 mL of concentrated
        hydrochloric acid, and then add 0.3 mL of 10% iron (III) chloride solution)
Acetylacetone Reagent (2 mL of acetylacetone dissolved in 98 mL of 1.0 M sodium
        carbonate; must be prepared fresh daily)
95% Ethanol
*p*-dimethylaminobenzaldehyde Reagent (677.5 mg *p*-dimethylaminobenzaldehyde in 25
        ml of 1:1 ethanol:HCl; must be prepared fresh daily)
Phenylhydrazine Reagent (neutralize 3 mL of phenylhydrazine with 9 mL of acetic acid in
        a 50 mL Erlenmeyer flask, add 15 mL of water, transfer mixture to a graduated cylinder,
        and add water to 30 mL)
0.5% Solution of red tetrazolium (0.5 g red tetrazolium in 100 mL of water)
Seliwanoff test (50 mg of rercinol dissolved in 33 mL of concentrated hydrochloric acid
        and then diluted with water to 100 mL)

## Experimental Procedure

### Barfoed's Test

1) Place 1 mL of a 1% solution of carbohydrate in a test tube. (Use 1 mL of water as a negative control.)
2) Add 0.5 mL of Barfoed's reagent to each tube.
3) Heat the tubes in a hot water bath. Note the time of formation of the red precipitate.

### Benedict's Test

1) Place 500 µL of a 1% solution of carbohydrate in a test tube. (Use 500 µL of water as a negative control.)
2) Add 2 mL of Benedict's reagent.
3) Place tubes in a boiling water bath for 2-3 minutes.
4) Remove the test tubes from the heat and carefully note the color of the precipitate (not the color of the solution).

### Bial Test

1) Place 1 mL of 1% solution of carbohydrate in a test tube. Add 1 mL of Bial's reagent.
2) Repeat Step 1 with 1 mL of water as a control.
3) Heat each tube in a hot water bath. Note the color of the solution.

### Elson-Morgan Determination of Hexosamines

1) Place 0.3 mL of 1% solution of carbohydrate in a screw-top test tube.
2) Repeat Step 1 with 1 mL of water as a control.
3) To each sample above, add 1 mL of acetylacetone reagent. Cap tubes and heat in a 90°C water bath for 45 minutes.
4) Cool the tubes in a bath of tap water; then add 4 mL of 95% ethanol. Add 1 mL of *p*-dimethylaminobenzaldehyde reagent. Mix the samples and observe the color.

**NOTE** – quantitive measurements can be made by creating a dilution series of N-acetylglucosamine standards containing 0.5µmol–2.5 µmol. Follow Steps 1-4, allow the solutions to incubate at room temperature for 1 hour. Determine absorbance of each solution at 540 nm. Use the standard curve of absorbance at 540 nm versus concentration of N-acetylglucosamine to determine the concentration of hexosamine in an unknown sample.

### Osazone Formation

1) Place 0.33 mL of phenylhydrazine reagent into a test tube.
2) Add 1 mL of 1% solution of carbohydrate.
3) Heat tubes in a hot water bath for 20 minutes, shaking tubes occasionally during heating.
4) Note the time at which the osazone product crystallizes from the mixture (mixture may appear cloudy).
5) After 20 minutes, cool to room temperature, and then place in an ice bath. If necessary, scratch the side of the test tube to initiate crystallization.
6) Note the time of formation and appearance of each of the precipitates.

*Red Tetrazolium Test*

1) Place one drop of a 1% solution of carbohydrate in a test tube. Add 1 mL of 0.5% solution of red tetrazolium and one drop of 3 M sodium hydroxide.
2) Heat the test tube in a beaker of hot water. Note the time of formation of a colored pigment.
3) A positive result occurs when an intensely colored pigment is formed.

*Seliwanoff Test*

1) Place 500 µL of a 1% solution of carbohydrate in a test tube. Use 500 µL of water as a control.
2) Add 2 mL of Seliwanoff reagent to each tube.
3) Put the tubes in a hot water bath. Note the time at which a color change occurs in the tubes. If after 10 minutes no change has occurred, you can assume a negative result.

## Optional Additional Explorations of Carbohydrate Identification

Design experiments to investigate the effects of interferences, such as buffer components, other macromolecules, etc., on the carbohydrate assays described above. Identify the carbohydrates present in food or naturally occurring substances, such as fruit samples, corn syrup, honey, or plant hydrolysates. Prior to any investigations, seek approval for the procedures from your instructor.

## Questions

1) Provide a flow chart showing the protocol used to identify your unknown. Provide an explanation of each chemical feature that is analyzed by the assays used.

2) What is the identity of your unknown compound?

3) What substances might interfere with these asssays? Could you use these assays to analyze for carbohydrates in the presence of other biomolecules?

## References for the Experiment

Bell, C. E., Taber, D. F., and A. K. Clark. *Organic Chemistry Laboratory with Qualitative Analysis.* 3rd edition. Harcourt College Publishers, 2001. pp. 315-321.

Clark, John M., and Robert L. Switzer. *Experimental Biochemistry.* 2nd edition. New York: W.H. Freeman and Company, 1977.

Malherbe, J. S., and C. J. Meyer. A Mini-Qualitative Carbohydrate Analysis Session. *J. Chem. Ed.,* 74, 1997: 1304-1305.

Schoffstall, A. M., Gaddis, B. A., and M. L. Druelinger. *Microscale and Miniscale Organic Chemistry Laboratory Experiments.* McGraw-Hill Higher Education, 2000. pp. 467-476.

Senkbeil, E. G. Inquiry-Based Approach to a Carbohydrate Analysis Experiment. *J. Chem. Ed.,* 76, 1999: 80-81.

Williamson, K. L. *Macroscale and Microscale Organic Experiments.* 3rd edition. Houghton Mifflin Company, 1999. pp. 672-678.

# Experiment 9. NMR Spectroscopy of Biological Molecules

NMR spectroscopy is becoming a valuable tool for biochemists. Some undergraduate institutions lack the resources to purchase and maintain a high-field NMR spectrometer. This experiment is composed of different exercises that could be conducted, depending on the type of instrument available. For more details on interpretation of NMR, please consult other textbooks on NMR spectroscopy.

## Prelaboratory Exercise

1) Draw the structures glucose, fructose, and sucrose in aqueous solution.

2) Predict the number of peaks that you would expect for each on a $^1$H and $^{13}$C NMR spectra.

## Characteristic of Carbohydrates Using 1H and 13C NMR

1) Create saturated solutions of glucose, fructose, and sucrose in 50% water/50% $D_2O$.

2) Obtain $^1$H and $^{13}$C spectra of each solution.

3) If possible, obtain DEPT, COSY, and HETCOR spectra of each sample.

## Questions

1) Based on their structures, how many carbon atoms do each of the sugars used in the experiment contain?

2) How many signals for distinct carbons are observed in the $^{13}$CNMR spectra for each sugar? Does this number agree with the number of carbons expected in each compound? If the numbers do not agree, explain your observations. (HINT: Would anomers have any effect on the spectroscopy properties?)

3) Comment on the use of each of these NMR spectroscopy techniques. Recent literature includes articles about the use of $^1$H NMR to monitor the carbohydrate content in beverages such as beer. Based on your findings, do you think NMR is a good analytical technique for quantitative and/or qualitative studies of carbohydrates?

## Monitoring Metabolic Pathways Using NMR

Glycolysis is an important metabolic pathway. Information about metabolism and kinetics of metabolic reactions can be monitored using NMR. This experiment will enable you to observe the anaerobic metabolism of glucose by yeast.

## Materials

*Saccharomyces cerevisiae* yeast suspension
[1-$^{13}$C]glucose stock solution
NMR tube

1) Place 400 μL of glucose stock solution in an NMR tube.

2) Obtain a $^{13}$C NMR spectrum of the glucose sample.

3) Add 200 μL of yeast suspension to the NMR tube.

4) Obtain $^{13}$C NMR spectra every 5-15 minutes.

5) Identify the carbons responsible for the peaks observed and the changes indicated over time.

6) Optional Analysis: Determine the rate of appearance of ethanol and/or glycerol or the rate of disappearance of glucose.

Optional additional experiments that students could conduct include
- the effect of temperature
- the effects of varying concentration of glucose
- the use of isotopically labeled glucose with the label on carbons other than C-1
- the effect of the amount of yeast
- the effects of inhibitors such as EDTA

## Questions

1) Why was isotopically labeled glucose used in the experiment? Could the same experiment be conducted with unlabeled glucose? What results would be obtained if the label was present on a different carbon of glucose?

2) Is there a difference in reactivity between the α and β anomers of glucose? Explain your answer in light of your experimental findings.

3) What are the intermediates and final product(s) that contain the $^{13}$C label? Does the NMR results indicate the formation of these molecules in glycolysis?

## References for the Experiment

Den Hollander, J. A., et al. $^{13}$C Nuclear Magnetic Resonance Studies of Anaerobic Glycolysis in Suspensions of Yeast Cells. *Proc. Nat. Acad. Sci.*, 76, 1979: 6096-6100.

Mega, T. L., Carlson, C. B., and D. A. Cleary. Following Glycolysis using $^{13}$CNMR. *J. Chem. Educ.*, 74, 1997: 1474-1476.

# Chapter 3 Optional Project

In this experiment, you will be provided with an unknown biological sample that could contain protein, DNA, and/or carbohydrate. You will need to determine the composition of your sample using spectroscopic techniques.

Using the procedures described in Chapter 3 and/or in the literature, design an experiment that will determine the composition of your sample. You can request samples of known composition to use as controls.

Final Report: The discussion in your report should include
1) composition of your sample

2) details of the procedures followed

3) a discussion of the problems encountered in determination of the composition of the sample (i.e., were there interferences in the methods utilized?)

4) proposal of other experiments that could have been conducted to assist in your analysis

# CHAPTER 4

# Enzyme Kinetics

A catalyst is a substance that increases the rate of a reaction without being changed in the overall process. Biological catalysts, enzymes, are typically proteins and are required to ensure that reactions in cells occur at a useful rate under physiological conditions. Cells are complex structures that contain numerous molecules that can react together. Enzymes have specific binding sites and only catalyze the reaction between desired compounds. In this way, substances react along specified pathways, and wasteful side reactions are minimized.

A chemical reaction is often theoretically described using a reaction coordinate, as shown in Figure 4.1. The reactant molecules collide with each other, and partial bond breakage and formation occurs. This high-energy state where transitional bonds are present is often referred to as the transition state, or activated complex. The energy required to reach this stage of the reaction is called the activation energy barrier, and the difference in energy from reactants to the transition state can be measured as $\Delta G^{\circ \pm}$. The energy difference between the reactants and products is expressed as $\Delta G^\circ$, dictates the likelihood that a reaction will occur, and is related to the position of the reaction, i.e., equilibrium constant. As shown in Figure 4.1, a catalyst lowers the activation energy of the reaction by proceeding through a lower-energy transition state. Catalysts affect the rate of the reaction but do not affect the thermodynamic possibility that the reaction will occur (i.e., $\Delta G^\circ$ is the same for both catalyzed and uncatalyzed reactions).

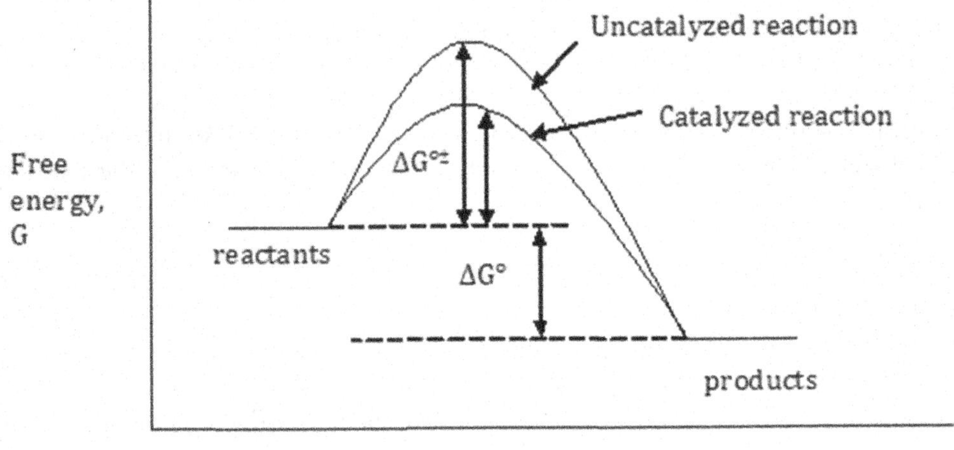


Figure 4.1 – A schematic diagram for an uncatalyzed versus catalyzed chemical reaction.

## 4.1 REACTION RATES

Kinetics is the study of rate changes associated with reactions. Consider the simple reaction:

$$A \longrightarrow P$$

The velocity ($v$) or rate of this reaction can be defined as the decrease in concentration of A over a period of time or the increase in concentration of P over a period of time:

$$v = \frac{-d[A]}{dt} = \frac{d[P]}{dt}$$

As a reaction proceeds, the concentrations of A and P will change over time. Since the concentrations are changing, the velocity at different times in the reaction coordinate will change. Figure 4.2 shows a plot of the concentration of P versus time. If a line is drawn tangent to the curve, the slope of the line is equal to the velocity at that time point ($v_t$). Many kinetic studies focus on the initial velocity ($v_0$)—the change in concentration in the first few seconds of the reaction.

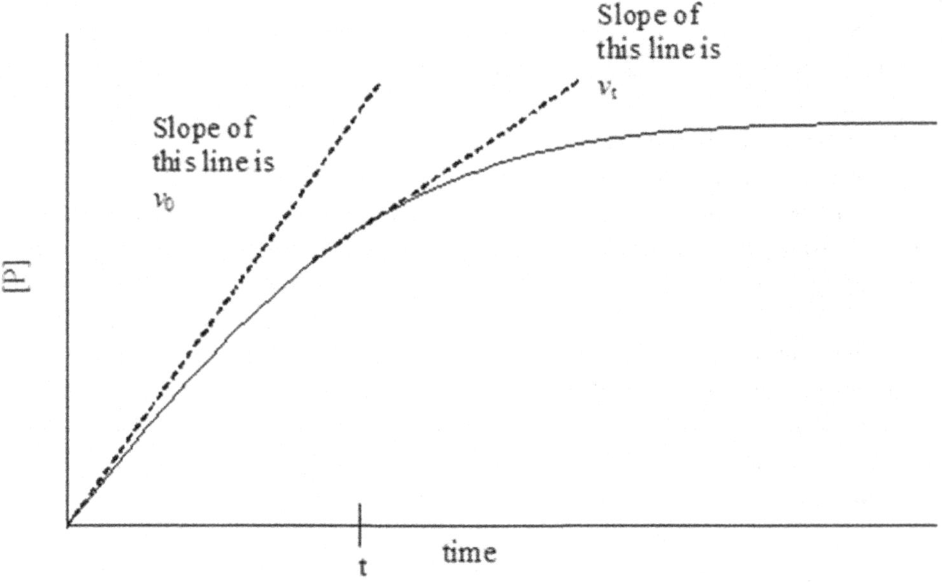


Figure 4.2 – A plot of the formation of product (P) versus time.
The velocity of the reaction can be determined from the slope of a line tangent to a curve.

The previously described rate expression provides average rate information over a specified time period. A differential rate law can be written to relate the rate of the reaction to the concentration of reactant:

$$v = \frac{-d[A]}{dt} = k[A]^m,$$

where   k  = rate constant
         m = order, usually an integer from 0 to 3

If a reaction involves more than one reactant (A and B), the rate of the reaction can be expressed as

$$v = \frac{-d[A]}{dt} = \frac{-d[B]}{dt} = k[A]^m[B]^n$$

The rate constant is a proportionality constant that relates concentration and rate. It has a fixed value at a specific temperature, but varies with temperature. The order (the exponents m and n) is not directly related to the stoichiometry of the reaction. It must be determined experimentally.

The overall order of a reaction is the sum of the exponents on each concentration term in the rate expression. The order indicates the dependence of the rate on concentration. In a first-order reaction (for example $v = k[A]$), the rate depends on the concentration of the reactant to the first power; therefore, doubling the concentration of A doubles the rate of the reaction. A second-order reaction may either depend on the concentration of a single reactant squared or on the concentrations of two reactants raised to the first power (i.e., $v = k[A]^2$ or $v = k[A][B]$). In the first case, doubling the concentration of A would quadruple the rate. In the second example, doubling only one concentration while holding the other constant doubles the rate, but if both the concentrations of A and B are doubled, the rate will quadruple. A zero-order reaction is one in which the rate of the reaction is independent of the concentration of the reactant(s).

Frequently, enzyme-catalyzed reactions initially exhibit first-order kinetics in that the rate of the reaction is directly dependent on the concentration of substrate. When the concentration of the reactant is large enough to saturate all of the binding sites on the enzyme, the enzyme is operating at maximum velocity, and the reaction exhibits zero-order kinetics. Once an enzyme is saturated, additional substrate will not increase the rate of the reaction. Figure 4.3 illustrates the dependence on rate of the concentration of substrate for an enzyme-catalyzed reaction.

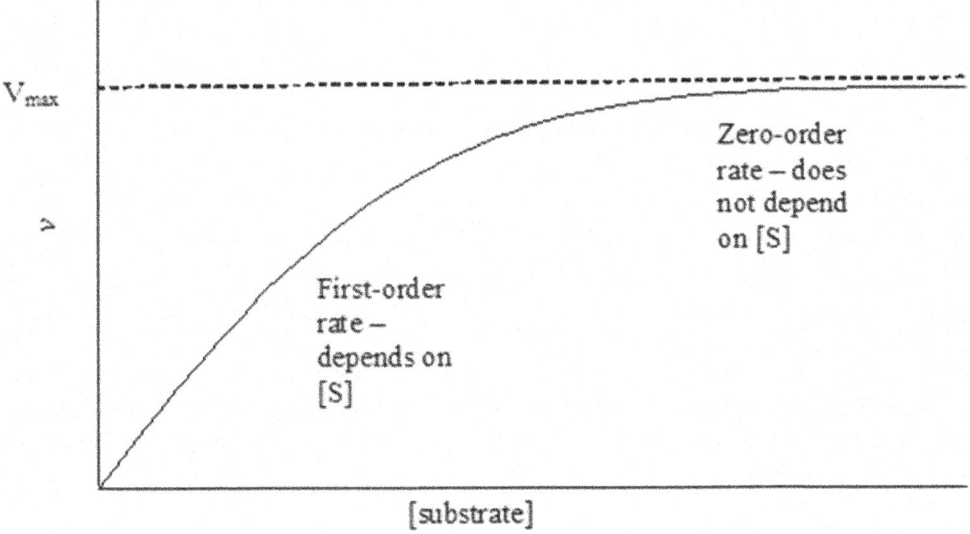


Figure 4.3 – The dependence on rate of concentration of substrate for an enzyme-catalyzed reaction.

The rate expressions previously presented provide information on the average rate and the dependence of rate on the concentration at a given time. Often, scientists want to know what the concentration of the reactant at a specific time is or the time required for the reaction to progress 50%. An integrated rate law, which is mathematically derived from the differential rate law, shows the relationship between concentration and time. The integrated rate law for a first order reaction is

$$\ln \frac{[A]_t}{[A]_0} = -kt, \text{ where}$$

$[A]_t$ = concentration of A at time t

$[A]_0$ = initial concentration of A

$t$  = time t

$k$  = rate constant

The half-life ($t_{1/2}$) of a reaction is defined as the time in which half of the original concentration of the reactants will be consumed. The half-life of a first-order reaction can be determined from

$$t_{1/2} = \frac{0.693}{k}$$

Second- and zero-order reactions exhibit different integrated rate laws and half-lives. For information about these equations, consult a general chemistry textbook.

## 4.2 MICHAELIS-MENTEN KINETICS

As shown in Figure 4.3, the rate of a reaction at a specific concentration of enzyme depends on the concentration of substrate. The reaction reaches a maximum velocity ($V_{max}$) under conditions of substrate saturation under specific reaction conditions (i.e., pH, temperature, etc.). $V_{max}$ reflects that the enzyme has specific binding sites for the substrate, and once saturated, the concentration of substrate will have no impact on the rate of the reaction.

In order for a reaction to occur, the enzyme (E) and substrate (S) must interact to form an ES complex before the product can be formed:

$$E + S \underset{k_{-1}}{\overset{k_1}{\rightleftharpoons}} ES \xrightarrow{k_2} E + P$$

Based on the reaction scheme above, enzyme-catalyzed reactions depend on the concentration of the enzyme and the substrate. A rate equation showing the correlation between velocity and concentrations of enzyme and substrate must be developed. In order to derive such an equation, several assumptions must be made:

1) The formation of product is assumed to be first-order and irreversible. This second step is considered the rate-determining step for the formation of product. Therefore, the rate of the reaction can be determined from

$$v_0 = \frac{d[P]}{dt} = k_2[ES]$$

   The problem with this equation is that it is extremely difficult (if not impossible) to measure the concentration of the transient ES complex.

2) The reaction is assumed to occur under steady-state conditions in which the concentration of ES complex is assumed to be constant. The rate of formation of ES is equal to $k_1[E][S]$. The rate of breakdown of ES is equal to $k_{-1}[ES] + k_2[ES]$. Under steady-state conditions, the rate of formation equals the rate of breakdown. Therefore, $k_1[E][S] = k_{-1}[ES]+k_2[ES]$. This can be rearranged to give an equilibrium expression ($K_m$):

$$K_m = \frac{k_{-1} + k_2}{k_1} = \frac{[E][S]}{[ES]}$$

3) Under saturating conditions (when the concentration of S is high), the entire enzyme is converted to ES. Therefore, it is assumed that the concentration of unbound or free enzyme is zero and the concentration of $E_{total}$ is equal to the concentration of ES.

4)  If all of the enzyme is in the form of ES, then the maximum velocity must be equal to $k_2[ES]$:

$$V_{max} = k_2[ES]$$

Using all of these assumptions and algebraic manipulations, the Michaelis-Menten equation was derived.

$$v_0 = \frac{V_{max}[S]}{K_m + [S]}$$

$V_{max}$ and $K_m$ are unique for individual enzymes under specific conditions of pH, temperature, and a specific concentration range of enzyme. The significance of each of these terms will be discussed in Section 4.5.

## 4.3  EFFECTS OF pH AND TEMPERATURE

Enzymes are proteins, and, as discussed in Chapter 2, proteins are affected by changes in pH. The effect of pH varies significantly for different enzymes. The addition of acid or base plays a role by affecting the ionization state of functional groups on the enzyme and substrate. Figure 4.4 shows an example of the effect of pH on the rate of two different enzymes. Enzyme 1 exhibits maximum activity around pH 3, whereas enzyme 2 is maximally active around pH 7. The difference in the effect of pH is due to the ionization state of the substrate and the amino acids in the binding and catalytic sites of the enzyme. If electrostatic interactions are essential for the formation of the ES complex and the addition of base converts a positively charged ammonium ion to the neutral amino group, the ionic interaction may cease to assist in the formation of the ES complex. At even higher pH levels, electrostatic repulsions may occur, which may lead to a decrease in interaction and activity.

Frequently, the optimal pH of an enzyme is related to the environment in which an enzyme is found. For example, pepsin assists in the digestion of proteins in the stomach and gut, a highly acidic environment. The pH optimum for pepsin is 1.5–2. Intracellular enzymes, such as phosphatase, have pH optimum close to the pH of a cell, between 7.5 and 8.

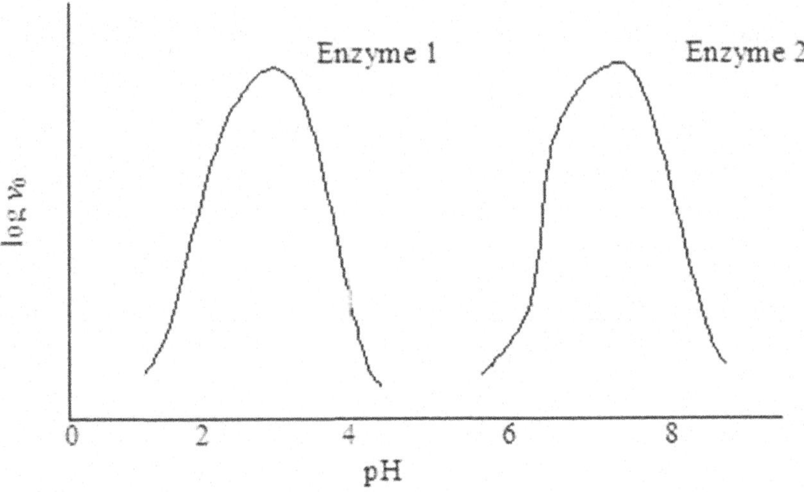


Figure 4.4 – Effect of pH on the rate of enzyme-catalyzed reactions.

Enzymes are also affected by temperature. Normally, the rate of a reaction increases as the temperature rises. This is due to an increase in molecular motions, which increase the likelihood of a collision and a resulting reaction. Also, at higher temperatures, the kinetic energy of a molecule increases, and this provides molecules with greater energy to overcome the activation energy barrier.

With biological reactions, however, an increase in temperature above 40-45°C results in a decrease in activity. As shown in Figure 4.5, the rate of an enzyme-catalyzed reaction will increase until an optimal temperature is reached. Above this temperature, the protein begins to denature and loses its native conformation. A denatured enzyme does not bind the substrate properly, and, often, the catalytic site is deformed such that the product cannot be produced from the substrate.

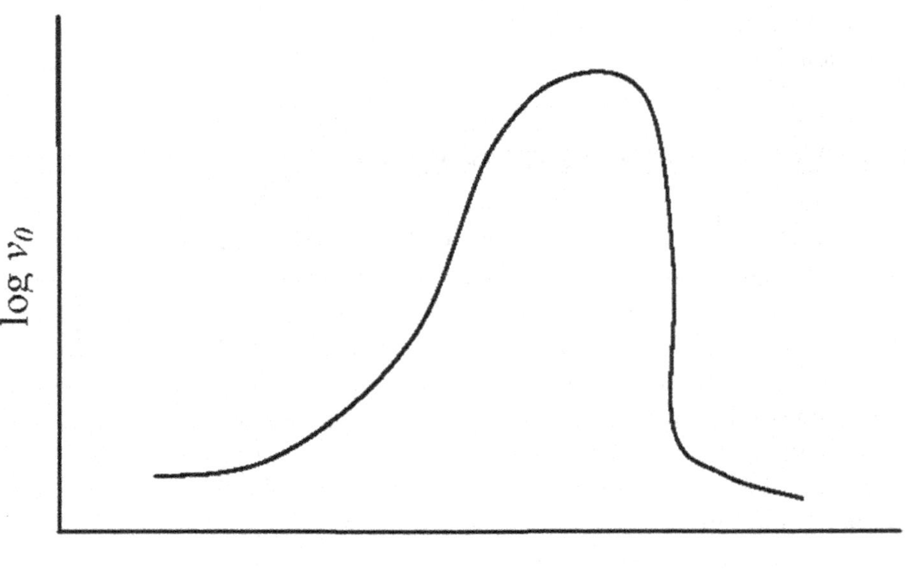


Figure 4.5 – Effect of temperature on the rate of an enzyme-catalyzed reaction.

## 4.4 GRAPHICAL TREATMENT OF KINETIC DATA

In order to illustrate the different types of graphical treatments of kinetic data, some hypothetical data will be used:

| [S] ($\mu$M) | $v_0$ (mM•s$^{-1}$) |
|---|---|
| 0.14 | 0.34 |
| 0.23 | 0.53 |
| 0.40 | 0.74 |
| 0.81 | 0.91 |
| 1.62 | 1.04 |

A plot of initial velocity versus concentration of substrate results in a hyperbolic curve. (See Figure 4.6.) $V_{max}$ can be determined from this plot by drawing an asymptote that approaches the curve at the point of enzyme saturation. $K_m$ can be determined from the plot by locating the concentration of substrate at the half-maximum velocity. For hand-drawn graphs, this method is not very accurate for determining $V_{max}$ and $K_m$. Recently, computer programs have been developed that conduct extensive nonlinear regressions that fit the data directly to the best hyperbola to determine $V_{max}$ and $K_m$. For the sample data, $V_{max}$ is approximated as 1.2 mM/s and the $K_m$ is 0.3 µM. CurveExpert (available at http://www.curveexpert.webhop.biz/) was used to perform the nonlinear regression of the data, and it was determined that the $V_{max}$ is 1.1(1.11) mM/s, and $K_m$ is 0.25 µM.

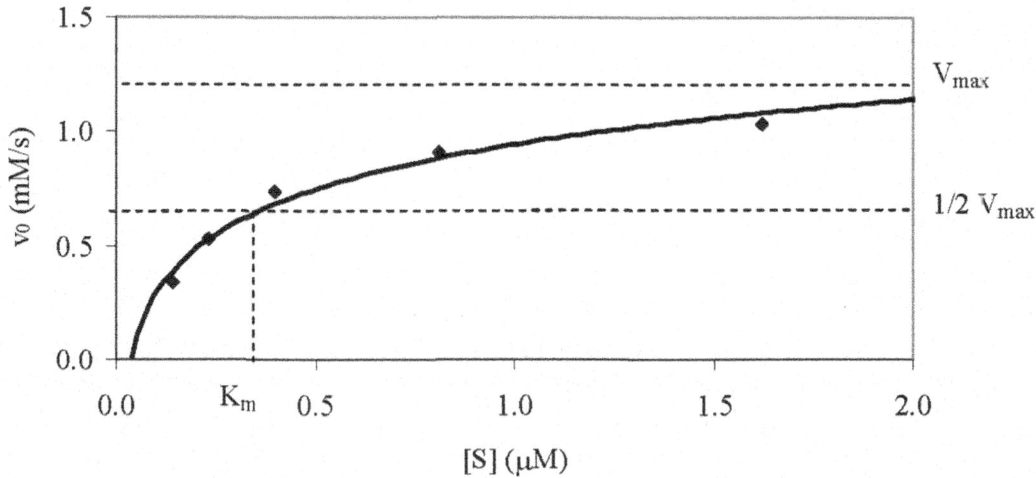


Figure 4.6 – Plot of initial rate of reactions versus concentration of substrate.

In 1934, Hans Lineweaver and Dean Burk published a paper in the *Journal of the American Chemical Society* describing a novel way to graph and visualize kinetic data. An interesting historical note was that the paper was rejected by all referees, but the editor decided to publish it anyway. This paper has become one of the most cited literature papers in history. Lineweaver and Burk manipulated the Michaelis-Menten equation to form a linear relationship between $1/v_0$ and $1/[S]$:

$$\frac{1}{v_0} = \frac{K_m}{V_{max}} x \frac{1}{[S]} + \frac{1}{V_{max}}$$

A plot of $1/v_0$ versus $1/[S]$ is linear with a slope that is equal to $K_m/V_{max}$, the x-intercept is equal to $-1/K_m$, and the y-intercept is equal to $1/V_{max}$. (See Figure 4.7.) This type of graphical interpretation of the data is called a Lineweaver-Burk, or double-reciprocal, plot. For the sample data, $K_m$ was determined to be 0.44 µM, and the $V_{max}$ was 1.5 (1.45) mM/s using the Lineweaver-Burk plot.

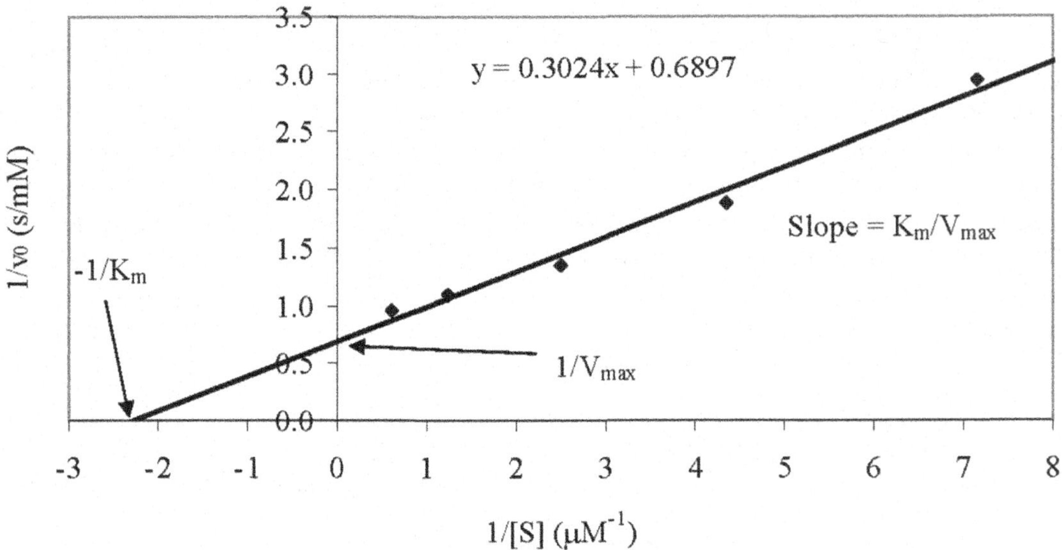


Figure 4.7 – Representative Lineweaver-Burk plot.

There are some disadvantages to using the Lineweaver-Burk plot to analyze enzyme data. Most experimental measurements involve higher concentrations of substrate, which compresses the data on the left side of the graph. Also, data at low concentrations of substrate are associated with a high degree of error; therefore, the 1/[S] values are associated with large errors. The compression of the data leads to an overemphasis of those data most connected with experimental error, and this leads to errors in the $K_m$ and $V_{max}$. Although the plot has been deemed by statistical analysis to be the less-acceptable means of enzyme kinetic analysis, it is still one of the most commonly used and recognized plots. These plots are also used to classify reversible enzyme inhibition. (See Section 4.6.)

Another way to display kinetic data graphically is to create an Eadie-Hofstee plot. (See Figure 4.8.) The Michaelis-Menten equation is rearranged as

$$v_0 = V_{max} - K_m \frac{v_o}{[S]}$$

A plot of $v_0$ versus $v_0/[S]$ gives a line with a slope equal to $-K_m$, an x-intercept equal to $V_{max}/K_m$, and a y-intercept equal to $V_{max}$. $K_m$ for the sample data is 0.35 μM, and $V_{max}$ is 1.3 (1.30) mM/s.

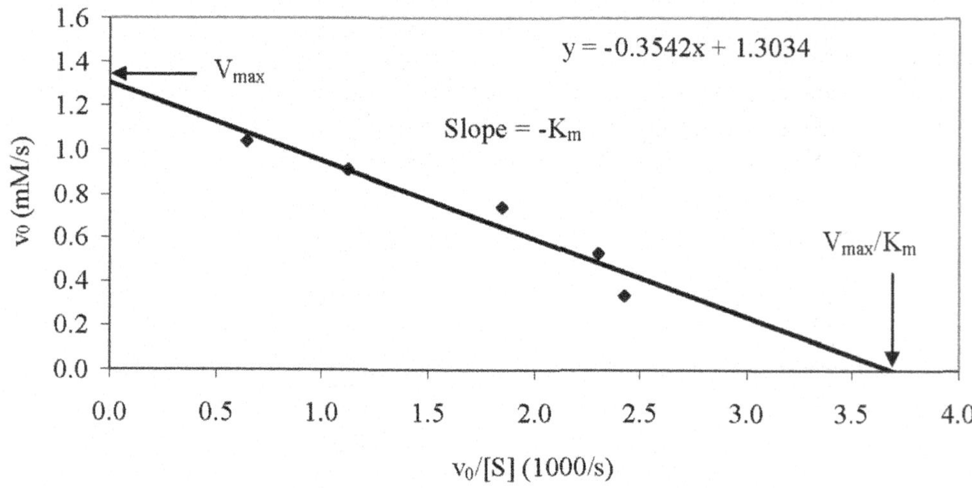


Figure 4.8 – A representative Eadie-Hofstee plot.

The Eadie-Hofstee plot is associated with less error than the Lineweaver-Burk plot. One problem with this type of analysis is that the initial velocity ($v_0$), which is a dependent variable, is plotted on both axes. An advantage of the Eadie-Hofstee plot is that outlying data points are often easily recognized.

The most accurate way to rearrange the Michaelis-Menten equation to graph enzyme kinetics data is a Haynes-Woolf plot. (See Figure 4.9.) The Michaelis-Menten equation is rearranged as

$$\frac{[S]}{v_0} = [S]\left(\frac{1}{V_{max}}\right) + \frac{K_m}{V_{max}}$$

A plot of $[S]/v_0$ versus $[S]$ gives a line with a slope equal to $1/V_{max}$, an x-intercept equal to $-K_m$, and a y-intercept equal to $-K_m/V_{max}$. $K_m$ for the sample data is 0.32 μM and $V_{max}$ is 1.3 (1.26) mM/s.

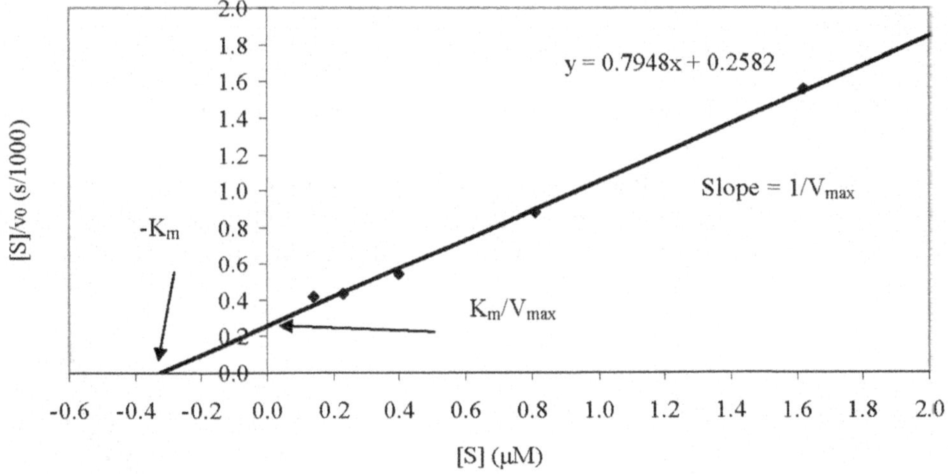


Figure 4.9 – Representative Haynes-Woolf plot.

A final method of kinetic analysis of enzyme data is the Eisenthal-Cornish-Bowden Direct Plot (also called Direct Linear Plot). (See Figure 4.10.) Each $v_0$, $[S]$ data set is plotted as the following (x,y) points: $(0, v_0)$ and $(-[S], 0)$. A line is drawn connecting the two points and extrapolated beyond the data points. For example, using the first data set, a line is constructed through (0, 0.34) and (-0.14, 0). The remaining $v_0$, $[S]$ data points are plotted in the same manner, and similar lines are created. The point (or points) of intersection of the lines are determined. The value at which the point of intersection crosses the x-axis is $K_m$, and the place where the line cross the y-axis is $V_{max}$. The Eisenthal-Cornish-Bowden direct plots are the most recommended plots when nonlinear curve-fitting programs are unavailable.

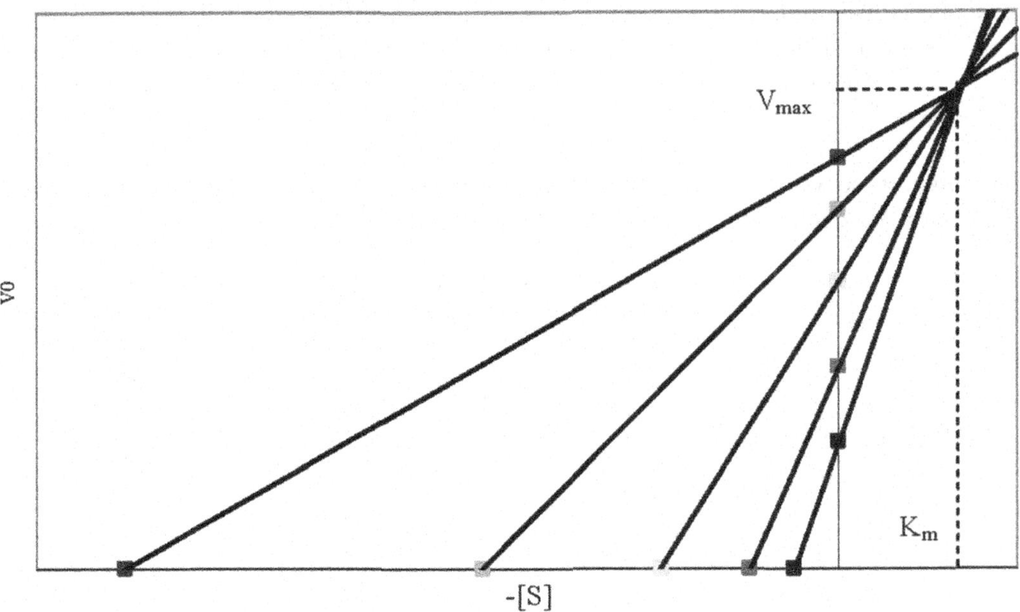


Figure 4.10 – Example of an Eisenthal-Cornish-Bowden direct plot.

Most kinetic data will be associated with error, and the lines will not intersect as neatly as shown in Figure 4.10. Often the data (as in the sample data used in this chapter) will have experimental error, and the lines will intersect at several points instead of a single point. (See Figure 4.11.)

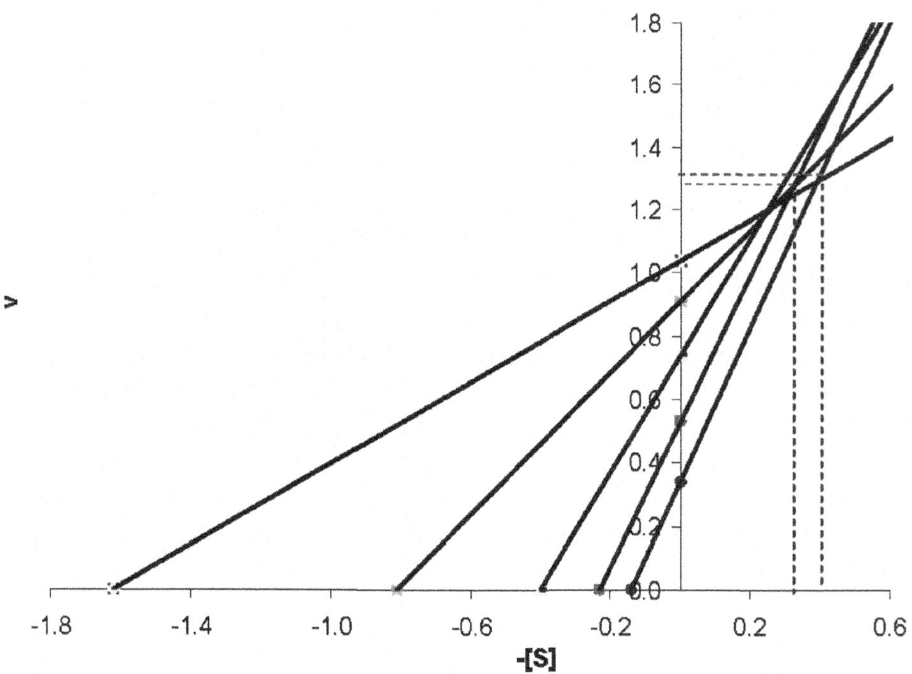


Figure 4.11 - Eisenthal-Cornish-Bowden direct plot with sample data.

If there is more than one intersection, the best way to estimate $K_m$ and $V_{max}$ is to determine the median value rather than determining all of the intersections and taking the mean of these results. The median value is the central value in a set of data arranged from highest to lowest.

Determining the median intersection will avoid the need to calculate or determine all of the intersections, which can exceed forty. It will also eliminate the influence of outliers, which can often occur if any of the lines are nearly parallel (as in Figure 4.11). In order to determine the median intersection, the number of intersections must be determined using the following equation:

$$\text{number of intersections} = \frac{n(n-1)}{2},$$

where   $n$ = number of lines

In the example in Figure 4.11, there are five lines, so there should be 10 intersections. The midway point in 10 measurements is between the fifth and the sixth data points. Starting at the bottom of the graph, proceed upward, counting the intersections until you reach the midway point (in this case, the fifth and sixth data points). Determine the y-axis coordinate at these points, and average the values to determine the $V_{max}$. Repeat the same procedure, moving left to right to determine the x-axis coordinate and the $K_m$. If three or more lines intersect in the same place, this intersection must be counted as the appropriate number of intersections in order to locate the median. Using the formula for determining the number of intersections, the number of crossings is calculated, and this is used to determine the median. In the example in Figure 4.11, the median intersections are indicated, and $K_m$ and $V_{max}$ are determined as 0.37 μM and 1.3 (1.27) mM/s, respectively.

The choice of kinetic analysis depends on the intent of the researcher. In order to determine the kinetic parameters accurately, a computer-assisted nonlinear curve fitting is the best choice. If this is not available, the recommended method of graphical analysis is the Eisenthal-Cornish-Bowden direct plot. For displaying kinetic behaviors or determining the type of enzyme inhibition, biochemists frequently use the Lineweaver-Burk plot. The main reason for the use of this plot is familiarity: it is the most commonly used and recognized kinetic plot.

# 4.5 SIGNIFICANCE OF $K_M$, $V_{MAX}$, AND $k_{CAT}$

Enzyme activity is frequently expressed as a unit (U), the amount (typically, μmoles) of substrate converted to product in a minute. The specific activity is the number of enzyme units per mg of protein that converts a specified amount of substrate to product in a time frame. Other measures of an enzyme's unique ability to catalyze the reaction are $K_m$, $V_{max}$, and $k_{cat}$.

There are numerous interpretations about $K_m$, the Michaelis-Menten constant. $K_m$ is frequently referred to as the affinity of an enzyme for the substrate. This is not totally accurate in that it neglects the contribution of $k_2$ to $K_m$. For enzymes where $k_2$ is significantly smaller than $k_{-1}$, $K_m$ can be considered the inverse of the enzyme-binding constant ($K_D$). Under these conditions, the greater the $K_m$, the lower the affinity the enzyme has for the substrate.

However, not all enzymes react in two steps in which $k_2$ is smaller than $k_{-1}$. Another definition of $K_m$ is the concentration of substrate that produces half-maximal reaction rate. This relationship can be derived by substituting $v_0 = \frac{1}{2} V_{max}$ into the Michaelis-Menten equation and solving for the concentration of substrate, which is equal to $K_m$. In this description, $K_m$ is a measure of the substrate concentration that is required for effective catalysis. For example, if an enzyme has a large $K_m$, a greater concentration of substrate will be needed to reach a given reaction velocity, as compared with one with a smaller $K_m$.

$V_{max}$ is the maximum rate that can be achieved by an enzyme at a given concentration. $V_{max}$ is achieved under saturation conditions and can be viewed mathematically as

$$V_{max} = k_{cat}[E]$$

$k_{cat}$ is often referred to as the enzyme turnover number, or the catalytic rate constant. This is a direct measure of the catalytic production of product under saturation/optimum conditions (the units of activity per mole of enzyme ($\mu$mol/min•mol enzyme)). The reciprocal of $k_{cat}$ is the time required for an enzyme to "turnover" or convert one substrate molecule to product.

At low substrate concentrations and high enzyme concentration, the most appropriate measure of enzymatic efficiency is $k_{cat}/K_m$. At low concentrations of substrate (a concentration in which the substrate is significantly less than that of $K_m$), the Michaelis-Menten equation can be used to derive the following equation:

$$v_0 \cong \frac{k_{cat}}{K_m}[E][S]$$

The term $k_{cat}/K_m$ can be considered as a second-order rate constant for the reaction between the enzyme and substrate. The larger the $k_{cat}/K_m$, the more efficient the enzyme. This is frequently the comparison used for different enzyme or substrate combinations.


## 4.6 ENZYME INHIBITION

Compounds that inhibit or alter the enzymatic reactions are useful for biochemists. Studies of enzyme inhibitors help scientists to learn how enzymes bind and interact with their substrates, to understand enzyme regulation in metabolism, to study the role of biological poisons and toxins, and to develop pharmaceutical agents to treat diseases. There are many different molecules that inhibit enzymatic activities. A major distinction between the classes of inhibitors is whether the effect or binding of the inhibitor is reversible (noncovalent) or irreversible (covalent). This section will focus on several classes of reversible, noncovalent inhibitors: competitive, noncompetitive, mixed, and uncompetitive.

Competitive inhibitors are molecules that bind to the same site as the substrate and therefore compete for the enzyme with the substrate. The following scheme illustrates the presence of a competitive inhibitor in the enzyme reaction pathway:

$$E + S \rightleftharpoons ES \longrightarrow E + P$$

$$+$$

$$I$$

$$\Big\updownarrow K_I$$

$$EI$$

Unlike the substrate, once the inhibitor is bound to the enzyme, it cannot be converted into a product. In order to reverse the inhibition, the competitive inhibitor must be removed from the binding site on the enzyme so the substrate can then bind. Because the inhibitor does not change the catalytic site of the enzyme, the maximum velocity can still be reached if sufficient substrate is present (i.e., the $V_{max}$ is unaffected). In order to reach the half-maximum velocity, a higher concentration of substrate is required, and $K_m$ is changed. It will take a higher concentration of substrate to reach the maximum velocity.

The presence of a competitive inhibitor influences the Michaelis-Menten equation:

$$v_0 = \frac{V_{max}[S]}{K_m\left(1+\dfrac{[I]}{K_I}\right)+[S]}$$

As described above, $V_{max}$ will remain the same for all concentrations of inhibitor. The apparent $K_m$ will be influenced by the concentration of the inhibitor and the enzyme affinity for the inhibitor ($K_I$). A Lineweaver-Burk plot of the kinetic data will demonstrate the effect of the competitive inhibitor on the enzyme. (See Figure 4.12.)

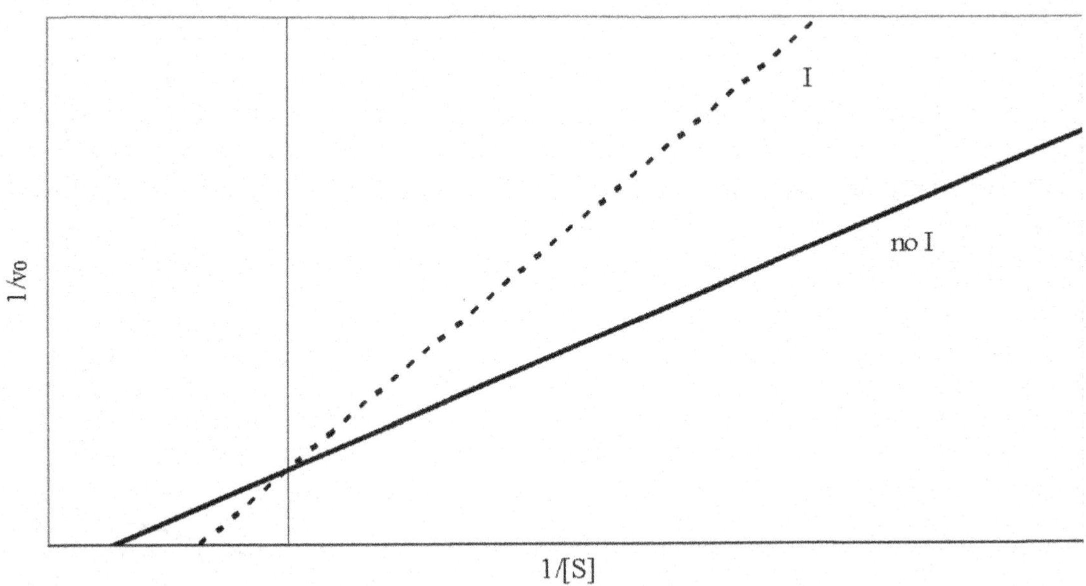


Figure 4.12 – Lineweaver-Burk plot for a hypothetic enzyme in the absence and presence of a competitive inhibitor (I).

Noncompetitive inhibitors are molecules that bind to a site other than the substrate binding site. The binding of the inhibitor to the enzyme reduces its catalytic activity. A noncompetitive inhibitor is capable of binding to both the enzyme and the enzyme substrate complex, as shown in the following scheme:

$$E + S \rightleftharpoons ES \longrightarrow E + P$$

Pure noncompetitive inhibitors are ones that have equivalent affinities for the enzyme and the enzyme substrate complex; therefore, $K_I$ and $K_I'$ are identical. If the affinities differ, the inhibitor is called a mixed inhibitor.

A noncompetitive inhibitor does not affect $K_m$ because it does not bind to the substrate binding site. However, the inhibitor alters the catalytic efficiency, and $V_{max}$ is reduced for all values of concentrations of substrate. The presence of a noncompetitive inhibitor influences the Michaelis-Menten equation:

$$v_0 = \frac{V_{max}[S]}{K_m\left(1 + \dfrac{[I]}{K_I}\right) + [S]\left(1 + \dfrac{[I]}{K'_I}\right)}$$

Lineweaver-Burk plots of the kinetic data demonstrate the effect of the pure noncompetitive and mixed inhibitors on the enzyme. (See Figures 4.13-4.15.)

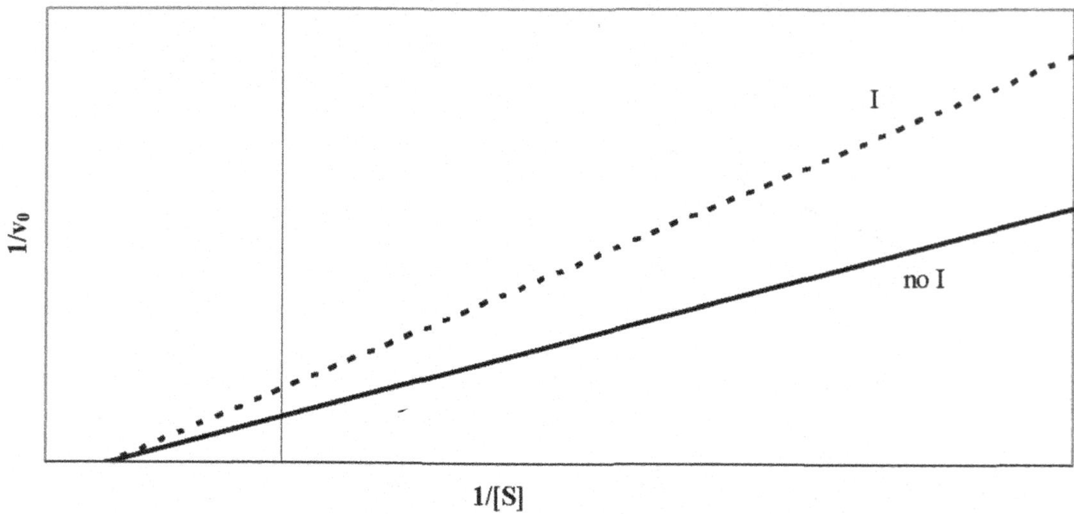


Figure 4.13 - Lineweaver-Burk plot for a hypothetic enzyme in the absence and presence of a pure noncompetitive inhibitor.

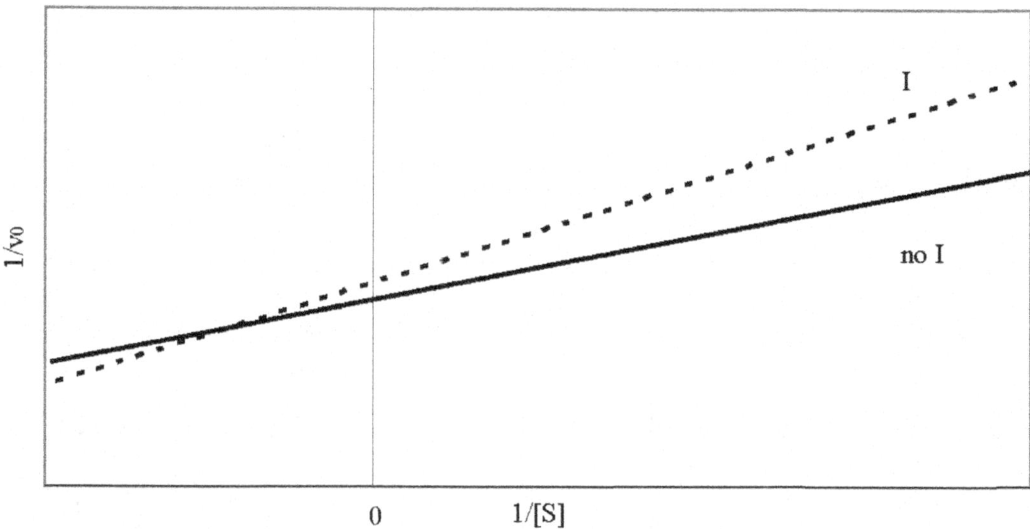


Figure 4.14 - Lineweaver-Burk plot for a hypothetic enzyme in the absence and presence of a mixed inhibitor for which $K_I$ is less than $K_I'$.

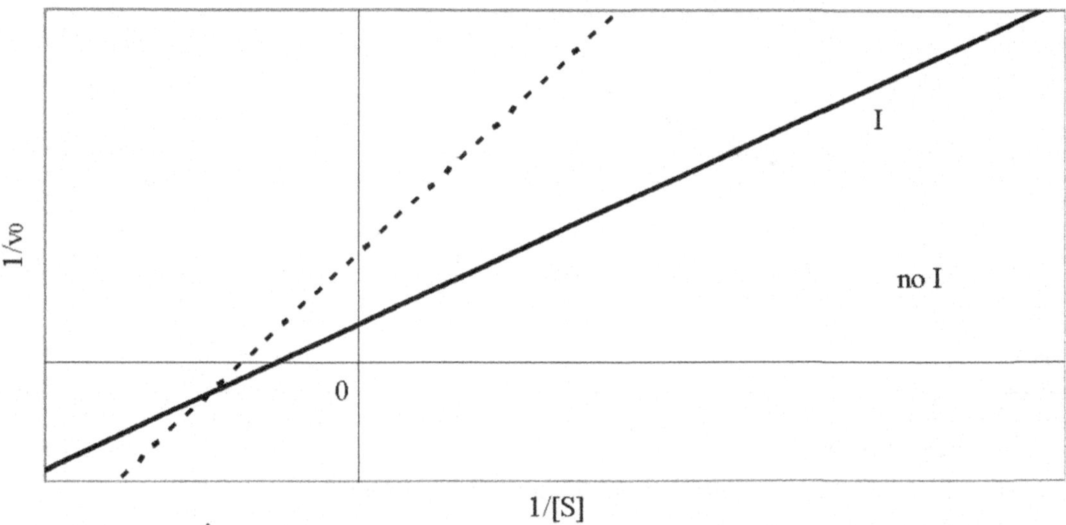


Figure 4.15 - Lineweaver-Burk plot for a hypothetic enzyme in the absence and presence of a mixed inhibitor for which $K_I$ is greater than $K_I'$.

The final class of reversible inhibitors describes inhibitors that act through an uncompetitive mechanism. An uncompetitive inhibitor only binds to the enzyme complex formation. The presence of the inhibitor prevents the conversion of substrate to product. The following reaction scheme illustrates the effect of an uncompetitive inhibitor:

$$
\begin{array}{ccc}
E + S \rightleftharpoons & ES & \longrightarrow E + P \\
& + & \\
& I & \\
& \updownarrow K_I & \\
& ESI &
\end{array}
$$

In the presence of an uncompetitive inhibitor, the Michaelis-Menten equation is altered to reflect the influence of the inhibitor:

$$v_0 = \frac{V_{max}[S]}{K_m\left(1+\dfrac{[I]}{K_I}\right)+[S]\left(1+\dfrac{[I]}{K'_I}\right)}$$

Figure 4.16 shows a Lineweaver-Burk plot in the absence and presence of an uncompetitive inhibitor.

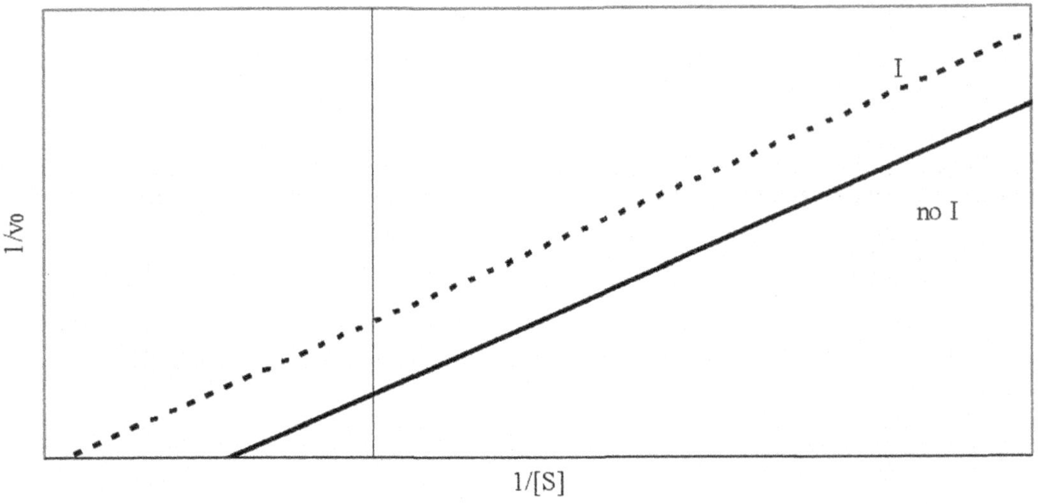


Figure 4.16 - Lineweaver-Burk plot for a hypothetic enzyme in the absence and presence of an uncompetitive inhibitor.

---

## Science in Action
### *Box 4.1 – Use of Enzyme Inhibitors as Clinical Drugs*
Most of the clinical agents used to treat disease that function by inhibiting enzymes act through a reversible mechanism of action. The majority of these agents act through competitive inhibition pathways, and examples of agents on the market include lovastatin, captopril, Viagra, and indinavir. There are several noncompetitive enzyme inhibitors in clinical use or trials, including Nevirapine, tacrine, and trazodone. Uncompetitive enzyme inhibitors are also on the market (Finasteride, mycophenolic acid, and camptothecin). Methotrexate is an interesting agent because it acts as a competitive inhibitor with regard to the substrate dihydrofolate and as an uncompetitive inhibitor with regard to NADPH.

---

This chapter briefly introduces enzyme kinetics and means of analysis. For multisubstrate systems, the topic is more complicated than described here. This chapter is meant to provide a initial understanding of enzyme kinetics. For additional information, see References and Further Reading.

# REFERENCES AND FURTHER READING

Boyer, R. *Modern Experimental Biochemistry*. 3rd Edition. San Francisco, CA: Benjamin Cummings, 2000.

Copeland, R. A. *Evaluation of Enzyme Inhibitors in Drug Discovery: A Guide for Medicinal Chemists and Pharmacologists. Methods of Biochemical Analysis,* Vol. 46. Hoboken, NJ: John Wiley and Sons, Inc., 2005.

Copeland, R. A. *Enzymes: A Practical Introduction to Structure, Mechanism, and Data Analysis*. New York, NY: John Wiley and Sons, Inc., 2000.

Corneley, K., et al. Kinetics of Papain. An Introductory Biochemistry Laboratory Experiment. *J. Chem. Ed.*, 76, 1999: 644-645.

Dagani, R. Straightening Out Enzyme Kinetics. *Chem. Engin. News,* 81, 2003: 27.

Dean, R. L. Kinetic Studies with Alkaline Phosphatase in the Presence and Absence of Inhibitors and Divalent Cations. *Biochem. Mol. Biol. Educ.,* 30, 2002: 401-407.

Farrell, S. O., and R.T. Ranallo. *Experiments in Biochemistry. A Hands-on Approach.* Florence, KY: Brooks Cole Thomas Learning, 2000.

Leskovac, V. *Comprehensive Enzyme Kinetics*. New York, NY: Kluwer Academic Publishers, 2003.

Marangoni, A. G. *Enzyme Kinetics: A Modern Approach.* Hoboken, NJ: John Wiley and Sons, Inc., 2003.

Nelson, D. L., and M. M. Cox. Lehninger Principles of Biochemistry. 4th Edition. New York, NY: W.H. Freeman and Co., 2005.

Sigma Chemical Product Information Bulletin, catalog number P5521.

**Websites**

Birch, P. *Online Enzyme Kinetics Course,* Department of Biological Sciences, University of Paisely. <http://orion1.paisley.ac.uk/kinetics/contents.html>.

*CurveExpert Website.* "A Curve Fitting System for Windows." <http://www.curveexpert.webhop.biz/ >.

# Experiment 10. Kinetic Studies of the Enzyme Papain

## Background

Papain is a cysteine endopeptidase (or thiol protease) derived from papaya (hence its name) and certain other plants. Papain (MW = 23.4 kDal) cleaves peptide bonds in proteins with broad specificity but prefers an amino acid with a large hydrophobic side chain at the P2 (amino terminal side of the cleavage site) position and does not accept Val in P1' (carboxy terminal side of the peptide bond) position. Other cysteine endopeptidases include bromelain from pineapple, ficin from figs, and actidin from kiwi.

Papain is commercially used to tenderize meat, to reduce viscosity and increase palatability in pet food, to clarify beer, to clean soft contact lenses, to remove hair from animal hides before tanning, and as a detergent to remove stains. It has been used medicinally to accelerate wound healing, to prevent cornea scar deformations, to treat jellyfish and insect stings, to treat edemas, and to treat indigestion and stomach ulcers.

Instead of using a protein as our substrate, we will use $N_\alpha$-benzoyl-arginine-$p$-nitroanilide (BAPNA). The structure contains an amide bond that resembles a peptide bond and is cleaved by papain to form a yellow compound $p$-nitroaniline.

$$\text{BAPNA} \xrightarrow[\text{H}_2\text{O}]{\text{papain}} + \quad p\text{-nitroaniline}$$

BAPNA is colorless, and the formation of $p$-nitroaniline can be followed spectrophometrically. **NOTE** – the absorbance of $p$-nitroaniline is pH dependent but does not change significantly over the range of pHs used in this experiment. A molar extinction coefficient of 10800 $M^{-1}m^{-}$ can be used in the analysis.

## Purpose of the Experiment

In this experiment, the progress of the reaction above will be monitored spectrophotometrically by measuring the increase in absorbance of $p$-nitroaniline ($\lambda_{max}$ 400 nm) with various concentrations of BAPNA and under various pH conditions. The kinetic properties of the enzyme papain (i.e., $K_m$, $V_{max}$, and pH optimum) will be determined.

## Prelaboratory Questions

1) Does the amount of enzyme present affect the kinetic data obtained in an experiment?

2) In the experiment, the formation of the *p*-nitroaniline is monitored for four minutes. Explain the indicated time frame for the experiment and why it was selected instead of a longer time, such as 10 minutes.

3) Determine the cellular location of papain and, based on this information, predict the pH optimum for papain.

## Materials

Papain extracts: dried papaya latex resuspended in 0.10 M phosphate buffer, pH 6,
       pH 7, and pH 8; in 0.10 M acetate buffer, pH 4 and pH 5; and in 0.10 M carbonate
       buffer, pH 9 and pH 10
0.020 M BAPNA in DMSO
DMSO
Phosphate buffers (0.1 M), pH 6, pH 7, and pH 8
Acetate buffers (0.1 M), pH 4 and pH 5
Carbonate buffers (0.1 M), pH 9 and pH 10
Cuvettes
Spectrophotometer
Bradford reagent (Coomassie Blue Solution)
Glass pipets

## Safety

Wear goggles. Dispose of the solutions at the end of the experiment in the proper manner, as indicated by the instructor.

## Experimental Procedure

*Part I - Preparation of BAPNA Solutions*

**NOTE –** <u>Do not use micropipetors to measure nonaqueous solutions. Use glass pipets to measure the BAPNA and DMSO.</u>

Create a dilution series of samples containing BAPNA in DMSO (0–0.020 M).

*Part II - Kinetic Studies of Papain*

1) Place 2.5 mL of 0.10 M phosphate buffer, pH 6, into two cuvettes. Set the spectrophotometer to 400 nm, and set the instrument to zero. (See Chapter 3.)

2) Remove the buffer from the sample cuvette. Place 2.50 mL of enzyme solution in 0.10 M phosphate buffer at pH 6 in the sample cuvette.

3) Add the volume of 0.50 mL of one of the BAPNA solutions prepared in Part I to each cuvette (sample and reference) at the same time. Mix the sample by covering the cuvette with parafilm and gently shaking. **Remember:** Do not use a micropipet to measure non-aqueous liquids. Use a microsyringe for the small volumes.

4) Place the cuvettes in the spectrophotometer. Record the absorbance at 400 nm every 15 seconds for 4 minutes.

5) Repeat Steps 1-4 using a different concentration of BAPNA in Step 3. Make sure to prepare a new sample and new blank each time.

6) If time permits, repeat the entire experiment.

*Part III —Determination of the pH-Activity Profile of Papain*

1) Place 2.5 mL of 0.10 M phosphate buffer, pH 6, into two cuvettes. Set the spectrophotometer to 400 nm and set the instrument to zero.

2) Remove the buffer from the sample cuvette. Place 2.5 mL of enzyme solution at pH 6 in the sample cuvette.

3) Add 0.50 mL of 0.020 M BAPNA solution to each cuvette (sample and reference) at the same time (note the time). Mix the samples by covering the cuvettes with parafilm and gently shaking. **Remember:** Do not use a micropipet to measure nonaqueous liquids. Use a microsyringe for the small volumes.

4) Place the cuvettes in the spectrophotometer. Record the absorbance at 400 nm every 15 seconds for 4 minutes.

5) Repeat Steps 1-4, substituting buffers at different pHs and enzyme solutions in the corresponding pH. Make sure to prepare a new sample and new blank each time. The blank should be the buffer in which the enzyme is dissolved.

6) If time permits, repeat the entire experiment.

*Kinetic Analysis of Data*

1) For each trial from Part II, create a plot of absorbance at 400 nm versus time (minutes). The slope of the line is equal to the initial velocity in absorbance units/minute. Using the extinction coefficient for *p*-nitroaniline, convert $v_0$ to units of M/min.

2) Create a Lineweaver-Burk plot, Haynes-Woolf plot, Eadie-Hofstee plot, and Eisenthal-Cornish-Bowden direct plot. If available, use a computer program to perform a nonlinear curve fit for a plot of initial velocity versus concentration of substrate.

3) Determine $K_m$ and $V_{max}$ using kinetic plots.

4) Create a pH-activity profile for the enzyme by plotting log $v_0$ versus pH.

## Questions

1) Compare the $K_M$ and $V_{max}$ for papain with $K_M$ and $V_{max}$ for other enzymes listed in your textbook. What is the significance of these values?

2) Compare the $K_M$ and $V_{max}$ determined from the different different plots (i.e., Lineweaver-Burk, Haynes-Woolf, etc.).

3) What is the pH optimum for papain? Briefly discuss the significance of this value.

## References

Corneley, K., et. al. Kinetics of Papain. An Introductory Biochemistry Laboratory Experiment. *J. Chem. Ed.*, 76, 1999: 644-645.

Nelson, D. L., and M. M. Cox. *Lehninger Principles of Biochemistry*. 3rd edition. New York, NY: Worth Publishers, 2000.

# Experiment 11. Analysis of Tyrosinase Enzymatic Activity

## Background

Tyrosinase, also called polyphenol oxidase, is a copper-containing enzyme that is found in both plant and animal cells. The enzyme has two catalytic activities: *o*-hydroxylation of monophenols to ortho diphenols, and aerobic oxidation of *o*-diphenols, such as catechol. Tyrosinase is a key enzyme in the biosynthesis of melanin in mammalian cells. In plant products, such as apples, potatoes, and bananas, the enzyme plays a role in the browning process occurring after injury.

One of the natural substrates of tyrosinase is 3,4-dihydroxy-L-phenylalanine (L-DOPA). The product of the reaction is dopachrome, which absorbs light at 475 nm. According to Fling, M. *et al.*, the molar extriction coefficient for dopachrome is 3600 $M^{-1}cm^{-1}$. In this experiment, we will monitor the enzymatic activity of tyrosine by monitoring the appearance of dopachrome at 475 nm.

The conversion of L-DOPA to dopachrome occurs in multiple steps. Figure 4.14 shows the pathway for the synthesis of dopachrome from L-DOPA. Tyrosinase is responsible for catalyzing the initial step. As the reaction is not a one-step process, the $K_M$ and $V_{max}$ for the enzyme are reported as apparent $K_M$ and $V_{max}$.

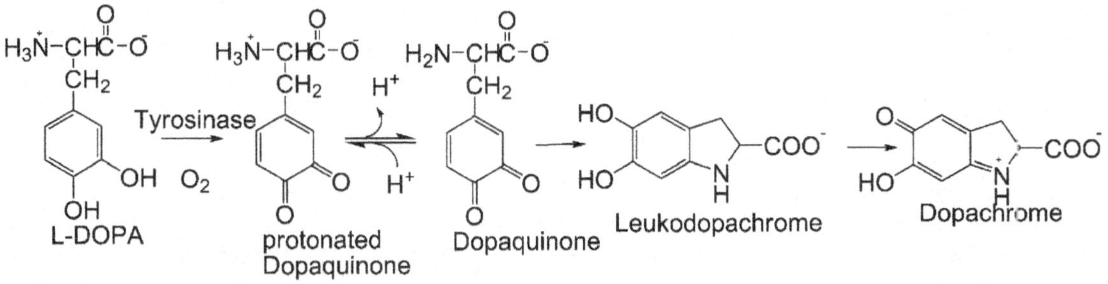

Figure 4.14 – The conversion of L-DOPA to dopachrome (from Seo, *et al.*).

## Purpose of the Experiment

In this experiment, the progress of the catalysis of the hydrolysis of L-DOPA will be monitored spectrophotometrically by measuring the increase in absorbance of dopachrome ($\lambda_{max}$ 475 nm) with various concentrations of L-DOPA and at varying pHs. The kinetic properties of the enzyme tyrosinase (i.e., apparent $K_m$, apparent $V_{max}$, and pH optimum) will be determined.

## Prelaboratory Questions

1) Does the amount of enzyme present affect the kinetic data obtained in an experiment?

2) In the experiment, the formation of the dopachrome is monitored for four minutes. Explain the indicated time frame for the experiment and why it was selected instead of a longer time, such as 10 minutes.

3) Determine the cellular location of papain and, based on this information, predict the pH optimum for papain.

## Materials

Tyrosinase (300 units/mL in 0.1 M phosphate buffer, pH 7)
L-DOPA (15 mM in the various buffer listed below, pH 3-8); prepare fresh – do not chill, as precipitation may occur
M phosphate buffer at pH 6, 7, and 8
0.1 M citrate buffer at pH 3, 4, and 5
Cuvettes
Spectrophotometer

## Safety

Wear goggles. Dispose of the solutions at the end of the experiment in the proper manner, as indicated by the instructor.

## Experimental Procedure

*pH Studies of Tyrosinase*

1) Place 470 μL of the citrate buffer at pH 3 and 500 μL of L-DOPA in the pH 3 buffer in two cuvettes.
   **NOTE** – if the cuvettes for the spectrophotometer must hold greater than 1 mL in order to record the absorbance, double the volume of each of the ingredients.

2) Mix the solutions in the cuvettes well and blank the spectrophotometer at 475 nm.

3) Add 30 μL of enzyme and quickly but gently mix the sample. Replace the cuvette in the spectrophotometer, and measure $A_{475}$ every 15 seconds for 4 minutes.

4) Repeat Steps 1-3 with the other buffers (pH 4, 5, 6, 7, and 8)

5) Identify the pH at which the enzyme produces the most dopachrome in 4 minutes.

*Kinetic Studies of Tyrosinase*

1) Place 470 µL of the buffer at the selected pH and 500 µL of L-DOPA in two cuvettes. **NOTE** – if the cuvettes for the spectrophotometer must hold greater than 1 mL in order to record the absorbance, double the volume of each of the ingredients.

2) Mix the solutions in the cuvettes well and blank the spectrophotometer at 475 nm.

3) Add 30 µL of enzyme and quickly but gently mix the sample. Replace the cuvette in the spectrophotometer, and measure $A_{475}$ every 15 seconds for 4 minutes.

4) Repeat Steps 1-3 with decreasing amounts of L-DOPA (from 7.5 mM to 0.25 mM). Make sure that the total volume in the cuvette, including the enzyme, is 1 mL. (See note with Step 1.)

5) If time permits, repeat the entire experiment.

6) If time permits, determine the concentration of protein in the enzyme sample using one of the methods in Experiment 5.

*Kinetic Analysis of Experimental Data*

1) For each trial from Kinetic Studies of Tyrosinase, create a plot of absorbance at 475 nm versus time (minutes). The slope of the line is equal to the initial velocity in absorbance units/minute. Use the extinction coefficient for dopachrome to convert $v_0$ to units of mM/min.

2) Create a Lineweaver-Burk plot, Haynes-Woolf plot, Eadie-Hofstee plot, and Eisenthal-Cornish-Bowden direct plot. If available, use a computer program to perform a nonlinear curve fit for a plot of initial velocity versus concentration of substrate.

3) Determine $K_m$ and $V_{max}$ using the kinetic plots.

4) For each trial from Determination of the pH-Activity Profile of Tyrosinase, create a plot of absorbance at 475 nm versus time (minutes). The slope of the line is equal to the initial velocity in absorbance units/minute. Use the extinction coefficient for dopachrome to convert $v_0$ to units of mM/min.

5) Create a pH-activity profile for the enzyme by plotting log $v_0$ versus pH.

6) If the amount of protein in the enzyme sample was determined, calculate the $k_{cat}$ for the enzyme.

## Optional Additional Explorations

Design experiments to investigate the effects of enzyme concentrations or temperature on the kinetic properties ($K_M$ and/or $V_{max}$) of the enzyme or the effects of stereochemistry on catalysis by investigating the effects of D-DOPA and L-DOPA as substrates of the enzyme. Prior to any investigations, seek approval for the procedures from your instructor.

## Questions

1)  Compare the $K_M$ and $V_{max}$ (and $k_{cat}$) for tyrosinase with values for other enzymes listed in your textbook. What is the significance of these values? How do your results compare with those in the literature for tyrosinase?

2)  Compare the $K_M$ and $V_{max}$ determined from the different plots (Lineweaver-Burk, Haynes-Woolf, etc). Is there a difference in the values determined? Which appears to be more accurate? Comment on your results.

3)  What is the pH optimum for tyrosinase? Briefly discuss the significance of this value.

## References for the Experiment

Fling, M., Horowitz, N. H., and S. F. Heinemann. The Isolation and Properties of Crystalline Tyrosinase from Neuropora. *J. Biol. Chem.,* 238, 1963: 2045-2053.

Rodriquez, M. O., and W. H. Flurkey. A Biochemistry Project to Study Mushroom Tyrosinase: Enzyme Localization, Isoenzymes, and Detergent Activation. *J. Chem. Educ.,* 69, 1992: 767.

Seo, S.-Y., Sharma, V. K, and N. Sharma. Mushroom Tyrosinase: Recent Prospects. *J. Agric. Food Chem.,* 51, 2003: 2837-2853.

Zhang, X., et al. Characterization of Tyrosinase from the Cap Flesh of Portabella Mushroom. *J. Agric. Food Chem.,* 47, 1999: 374-378.

# Experiment 12. Inhibition of Tyrosinase Activity

Enzymatic browning of foods is an aesthetics issue for consumers. One common practice to prevent browning of cut apples is to toss the fruit in a small amount of lemon or other citrus juice. Food chemists strive to find inhibitors of the enzyme tyrosinase in order to prevent the browning of vegetables and fruits. In this experiment, several tyrosinase inhibitors will be investigated.

## Purpose of the Experiment

For more information on tyrosinase, refer to Experiment 11. In this experiment, the inhibition of tyrosinase by *trans*-cinnamaldehyde, benzaldehyde, and benzoic acid will be investigated. This experiment can be conducted in conjunction with Experiment 11.

## Prelaboratory Question

Draw the structures of the inhibitors to be investigated and determine if they share any common features with the substrate of the enzyme. Based on the chemical structures, can you make any predictions about the type of inhibition that could be exhibited by the compounds?

## Materials

Tyrosinase (300 units/mL in 0.1 M phosphate buffer, pH 7)
L-DOPA (15 mM in 0.1 M phosphate buffer, pH 7); prepare fresh
Phosphate buffer at pH 7
10 mM *trans*-cinnamaldehyde
10 mM benzoic acid
Cuvettes
Spectrophotometer

## Safety

Wear goggles. Dispose of the solutions at the end of the experiment in the proper manner, as indicated by the instructor.

## Experimental Procedure

*Inhibitor Studies of Tyrosinase*

If you are conducting this experiment in conjunction with Experiment 11, trials without the inhibitors can be skipped, as they have already been completed.

1) Place 470 µL of the buffer at the selected pH and 500 µL of L-DOPA in two cuvettes. **NOTE** – if the cuvettes for the spectrophotometer must hold greater than 1 mL in order to record the absorbance, double the volume of each of the ingredients.

2) Mix the solutions in the cuvettes well and blank the spectrophotometer at 475 nm.

3)  Add 30 µL of enzyme and quickly but gently mix the sample.  Replace the cuvette in the spectrophotometer, and measure $A_{475}$ every 15 seconds for 4 minutes.

4)  Repeat Steps 1-3 with decreasing amounts of L-DOPA (7.5 mM to 0.25 mM).  Make sure that the total volume in the cuvette, including the enzyme, is 1 mL.  (See note with Step 1.)

5)  Repeat Steps 1-4 using the same concentrations of L-DOPA, but add 10 µL of the inhibitor solution to both cuvettes. Make sure that the total volume in the cuvette, including the enzyme, is 1 mL. (See note with Step 1.)

6)  Repeat Step 5 with the other inhibitor.

7)  If time permits, repeat the entire experiment at another concentration of inhibitor.

*Kinetic Analysis of Experimental Data*

1)  For each trial from Kinetic Studies of Tyrosinase, create a plot of absorbance at 475 nm versus time (minutes).  The slope of the line is equal to the initial velocity in absorbance units/minute. Using the extinction coefficient for dopachrome to convert $v_0$ to units of mM/min.

2)  Create a Lineweaver-Burk plot for the kinetic data in the presence and absence of each inhibitor.

3)  Determine apparent $K_M$ in the presence and absence of inhibitors, apparent $V_{max}$ in the presence and absence of inhibitors, and $K_I$ using the kinetic plots.

## Optional Additional Explorations

Design experiments to investigate the effects of other inhibitors of tyrosinase.  Prior to any investigations, seek approval for the procedures from your instructor.

## Questions

1)  Based on your findings, classify each of the inhibitors using the classes of reversible inhibitors presented in Chapter 4.

2)  Based on the classification of the inhibitors, what can you conclude about the binding site of the enzyme?

## References for the Experiment

Fling, M., Horowitz, N. H., and S. F. Heinemann. The Isolation and Properties of Crystalline Tyrosinase from Neuropora. *J. Biol. Chem.*, 238, 1963: 2045-2053.

Rodriquez, M. O., and W. H. Flurkey. A Biochemistry Project to Study Mushroom Tyrosinase: Enzyme Location, Isoenzymes, and Detergent. *J. Chem. Educ.*, 69, 1992: 767-769.

Seo, S.–Y., Sharma, V. K., and N. Sharma. Mushroom Tyrosinase: Recent Prospects. *J. Agric. Food Chem.*, 51, 2003: 2837-2853.

Zhang, X., et al. Characterization of Tyrosinase from the Cap Flesh of Portabella Mushroom. *J. Agric. Food Chem.*, 47, 1999: 374-378.

# Electrophoresis

5

## 5.1 THEORY OF ELECTROPHORESIS

Biological macromolecules can migrate in electrical fields according to their net charges or size due to a phenomenon known as electrophoresis. This technique was developed in the 1930s by Arne Tiselius, and he was awarded the Nobel Prize for this work in 1948. Molecules can be separated in the electrical field based on different mobilities, which are affected by charge, shape, size, and composition of the molecule. Separation techniques, other than electrophoresis, are described in Chapters 6 and 7. There are two types of electrophoresis: 1) solution electrophoresis, in which molecules separate in aqueous buffer in the absence of a solid or matrix support and 2) zone or slab electrophoresis, in which samples are applied in spots (zones) on a solid or matrix media in an aqueous ionic solution. The most common type of electrophoresis used in biological applications is zone electrophoresis, and this will be the focus of this chapter.

In zone electrophoresis, a slab or column of gel support material is placed in a buffered solution between two electrodes: the cathode, a negatively charged electrode; and the anode, a positively charged electrode. Positively charged molecules will travel toward the cathode, while negatively charged ions will travel toward the anode. The velocity of the molecule traveling in the media can be described by the following equation:

$$v = \frac{Eq}{f},$$

where   $v$ = velocity of the molecule
       E = applied voltage
       $q$ = net charge of molecule
       $f$ = frictional coefficient that depends on mass and shape of the molecule

As indicated in the equation, the migration of the molecule depends on the applied electric field ($E$), charge of the molecule ($q$), and size and shape of the molecule ($f$). Round proteins move faster than rod-shaped proteins. The movement of larger particles will be retarded in a dense media more than smaller-sized substances.

Often in biological electrophoresis applications, the voltage is held constant; however, constant current electrophoresis can be carried out as well. If the electrophoresis is carried out at constant voltage, the velocity of the particle depends on the ratio of $q$ to $f$. If the molecules have similar conformations (i.e., all linear pieces of DNA or spherical proteins), then the frictional coefficient will be based mainly on the mass of the molecule. Therefore, particles are separated based on a charge to mass ratio. If each particle has a unique charge to mass ratio, it will exhibit a unique mobility through the solid matrix, and separation will be achieved. If two particles have the same mass and shape, the one with the larger charge will move faster. If two molecules have the same mass and charge, the one with the more symmetrical (spherical) shape will move through the matrix faster.

The rate of separation by electrophoresis depends on the applied voltage. Rapid separation can be accomplished by increasing the voltage. However, this is not always advisable. Higher voltage causes increased heat, which can denature biological molecules or the matrix material. Better separation is achieved when a gel is run slowly, resulting in sharper, more-defined bands and less streaking.

## 5.2 TYPES OF GELS

There are two types of polymer materials that are generally used to create the gel matrix for zone electrophoresis. Agarose, a polysaccharide derived from agar, which is isolated from seaweed, is commonly used to separate nucleic acids. Polyacrylamide, a synthetic polymer of acrylamide, is often utilized in separations of nucleic acids or proteins. Each polymeric material will be briefly discussed in more detail.

### *Agarose Gels*

Agarose is a polysaccharide of galactose and 3,6-anhydrogalactose. The basic repeating unit of the polymer is shown in Figure 5.1. In solution, the polymer tends to form a left-handed helical structure that intertwines and aggregates to form a dense gel with pores ranging from 50 to >200 nm. Agarose is sold as the dried polymer, containing approximately 800 galactose residues per chain.

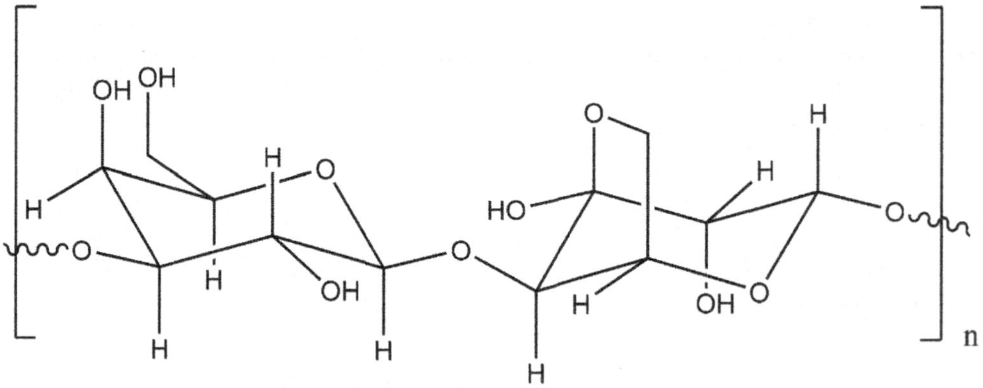

Figure 5.1 – Repeating unit in agarose.

Agarose is commercially available in several forms, and the properties of each type differ slightly. Standard (high-melting temperature) agaroses are isolated from various species of seaweed. These agaroses melt between 85–95°C and gel between 35–45°C (depending on manufacturer and species of seaweed used). Standard agarose is used to analyze and isolate DNA fragments up to 25 kilobases (kb) in length. Low-melting/gelling-temperature agaroses are chemically modified by hydroethylation. The chemical substitution lowers the temperatures at which the polymers melt and gel, typically between 40–65°C and between 8–35°C (depending on manufacturer and degree of modification), respectively. These agaroses are used for rapid purification of DNA fragments from gels.

Agarose gels are prepared by gently heating an appropriate quantity of the dried polymer in a buffer solution until the polysaccharide melts. Care must be taken not not boil the solution, as the concentration of agarose will change, and the sugar may burn or char. Once a homogenous solution is created, the warm solution (about 50–60°C) is poured into a gel cast, and a comb is used to create wells in which the samples will be applied. The gel is allowed to cool in order to harden.

Agarose gels are typically run horizontally in what is referred to as a submarine mode. (See Figure 5.2 for a picture of an example of a submarine gel electrophoresis apparatus and agarose gel.) The gel is immersed in a buffer solution, usually the same one in which the gel was cast. The buffered solution plays a number of roles. First, the buffer in the gel provides ions that will carry the current enabling the electrophoretic separation of the molecules. If the gel was cast in water, there would be no movement of molecules through the gel. Secondly, the buffer solution in which the gel is immersed helps to dissipate heat and eliminate some of the problems previously described.

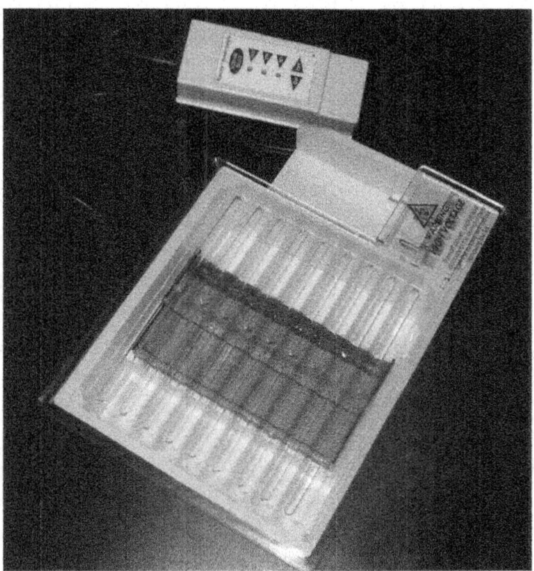

Figure 5.2 – An example of a submarine-style gel electrophoresis set-up. An agarose gel is in the gray tray in the center of the apparatus.

Agarose gels are run under native conditions, meaning that there is nothing present in the solutions or matrix that would denature the biological molecules. As mentioned above, agarose is more commonly used to separate nucleic acids. Table 5.1 shows the effective range of separation of DNA by agarose. These ranges will differ slightly depending on the type of agarose utilized.

Table 5.1 – Typical Range of Separation of DNA in Gels of Varying Concentrations of Standard Agarose.

| Agarose Concentration (% w/v) | Effective Range of Separation of Linear DNA (kb) |
|:---:|:---:|
| 0.3 | 5 – 60 |
| 0.5 | 2 – 25 |
| 0.7 | 0.8 – 10 |
| 0.9 | 0.5 – 7 |
| 1.2 | 0.4 – 6 |
| 1.5 | 0.2 – 3 |
| 2.0 | 0.1 – 2 |

### *Polyacrylamide Gels*

The matrix for polyacrylamide gel electrophoresis (PAGE) is polymers of acrylamide cross-linked with N,N'-methylene-bis-acrylamide. (See Figure 5.3.) The reaction is catalyzed by the formation of free radicals from the persulfate ion (supplied by ammonium persulfate, APS) N,N,N',N'-tetramethylethylenediamine (TEMED) and increases the rate of the reaction by enhancing the formation of radicals from APS. These radicals initiate the polymerization of acrylamide monomers. In the presence of the bis-acrylamide, the chains become cross-linked. The amounts of TEMED and APS will determine the rate of polymerization.

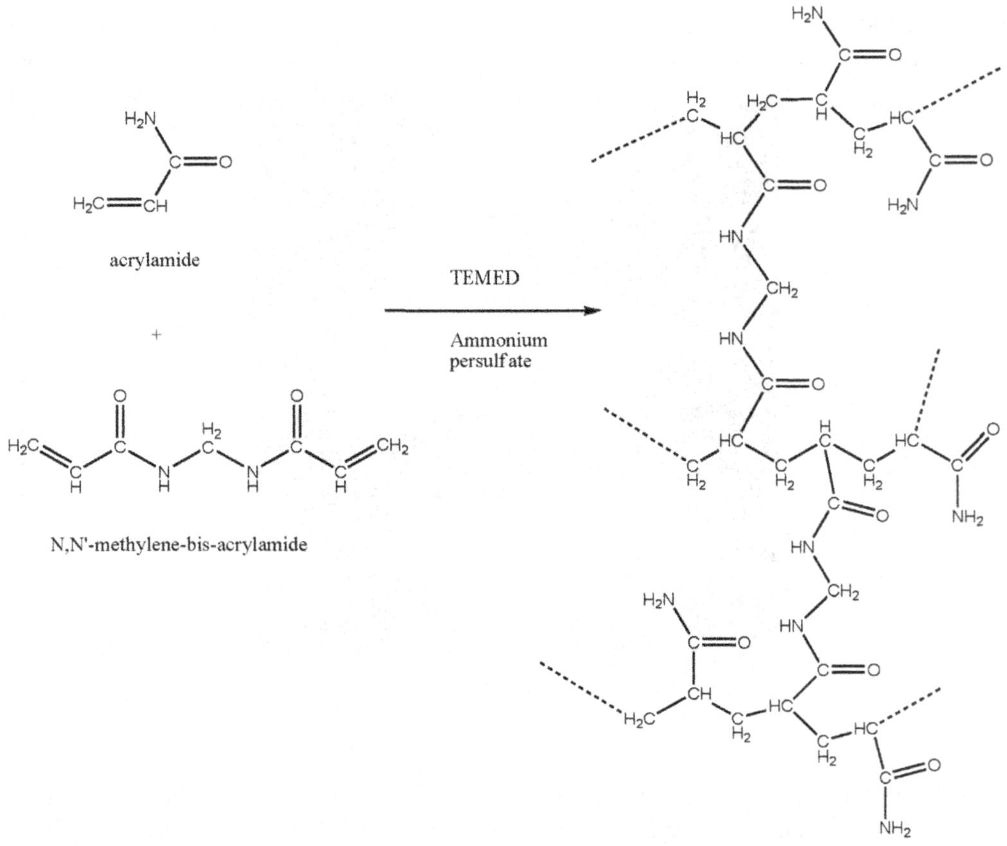

Figure 5.3 – Polymerization of acrylamide and bis-acrylamide into the polyacrylamide gel matrix.

The percentage of acrylamide and bis-acrylamide dictate the size of the pores in the gel support. A smaller concentration of the two molecules will result in larger pore sizes, creating a gel matrix that is appropriate for the separation of high-molecular-weight biomolecules. Likewise, a larger concentration of acrylamide and bis-acrylamide will result in a matrix with smaller pores that separates molecules of lower molecular weights. Tables 5.2 and 5.3 show the effective range of separations of DNA and proteins in polyacrylamide gels.

Table 5.2 – Effective Range of Separation of DNA by PAGE.

| Acrylamide (% w/v) (29:1 molar ratio of acrylamide to bis-acrylamide) | Range of Separation (bp) |
|---|---|
| 3.5 | 1000 – 2000 |
| 5.0 | 80 – 500 |
| 8.0 | 60 – 400 |
| 12.0 | 40 – 200 |
| 15.0 | 25 – 150 |
| 20.0 | 5 – 100 |

Table 5.3 – Effective Range of Separation of Proteins by PAGE.

| Acrylamide (% w/v) (29:1 molar ratio of acrylamide to bis-acrylamide) | Range of Separation (kDa) |
|---|---|
| 15 | 10 – 45 |
| 12.5 | 15 – 60 |
| 10 | 20 – 80 |
| 7.5 | 30 – 100 |
| 5 | 60 – 212 |

Polyacrylamide gels can be run either in the submarine fashion or in a vertical electrophoresis apparatus with the gel polymerized between two glass plates. The latter application is more commonly used in biochemistry laboratories. Figure 5.4 shows a vertical gel electrophoresis apparatus and a polyacrylamide gel.

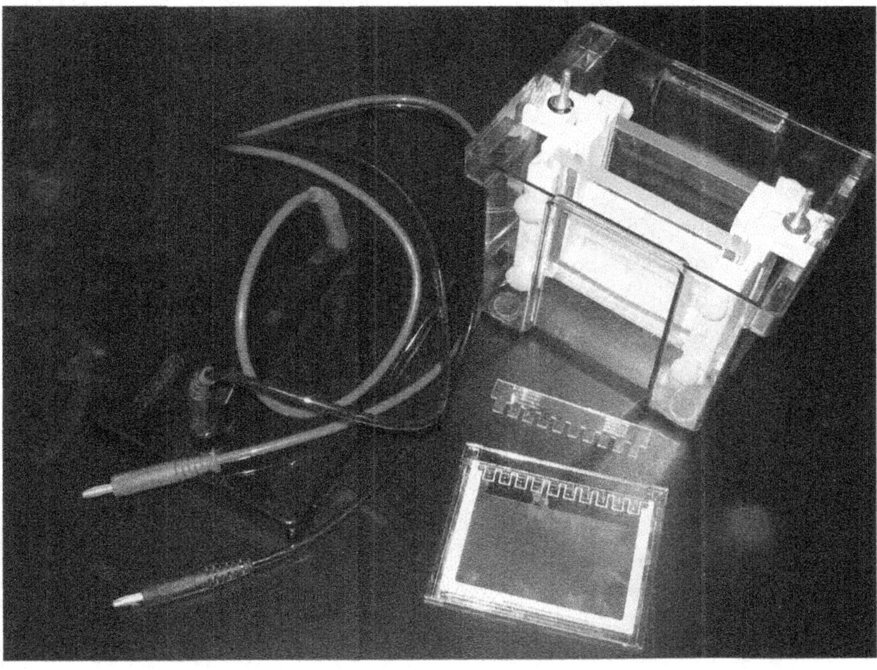

Figure 5.4 – Vertical electrophoresis apparatus and polyacrylamide gel.

## Discontinuous Gel Electrophoresis

Discontinuous gel electrophoresis utilizes two gels stacked on top of each other. The first, the stacking gel, is closest to the top, near the sample wells. This gel contains a lower concentration of acrylamide and is created with a lower-pH, lower-ionic-strength buffer. The role of this gel is to compress the molecules, usually proteins, into a compact band. The second layer, referred to as either the running, resolving, or separating gel, separates the proteins into discrete bands. This gel is made of a higher percentage of acrylamide in a higher-pH, higher-ionic-strength buffer.

The polyacrylamide gel slabs for discontinuous gel electrophoresis are typically prepared in buffer composed of Tris-HCl and glycine. Recall from Chapter 2 that glycine is one of the common amino acids and exists in several ionized forms. (See Figure 5.5.)

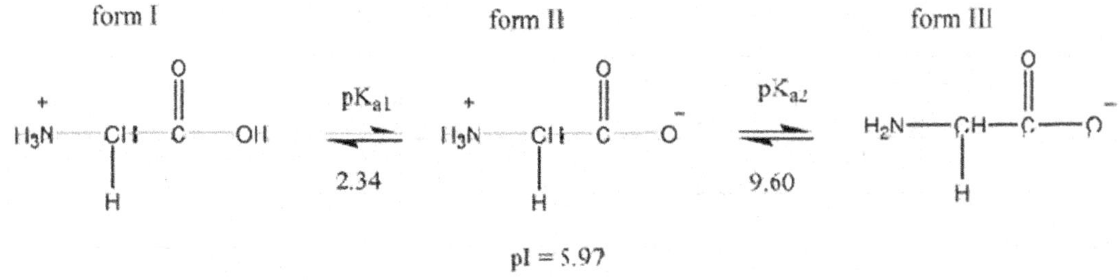

Figure 5.5 - Acid dissociation equilibria of glycine.

Figure 5.6 details the separation of proteins using discontinuous gel electrophoresis. Frequently, the pH of the stacking gel is 6.9, and the pH of the resolving gel is 8.9. In the stacking gel, the glycine will have a partial negative charge. The chloride from Tris-HCl is negatively charged and small in size and, therefore, has the largest $q/f$ ratio. At a pH of 6.9, most proteins will have a negative charge and a greater negative charge than that on glycine. (See Table 2.1 for $pK_a$s of amino acids.) Based on the charges and frictional coefficients, the relative velocity toward the anode would be glycine⁻ < proteins⁻ < Cl⁻. The result of the difference in mobility and the low percentage of acrylamide (thus, large pores) is the creation of a discrete band of proteins sandwiched between the glycine and the chloride ions.

The bands continue to move into the resolving gel. The chloride ions with a negative charge and small size will move quickly toward the anode. The glycine will be titrated at a pH of 8.9 to primarily form III. (See Figure 5.5.) This will increase the mobility of the glycine, and it will move past the proteins. The ionization state of the proteins will also change, and they will be separated in the higher concentration of acrylamide gel based on size, shape, and charge, as illustrated in (C) of Figure 5.6.

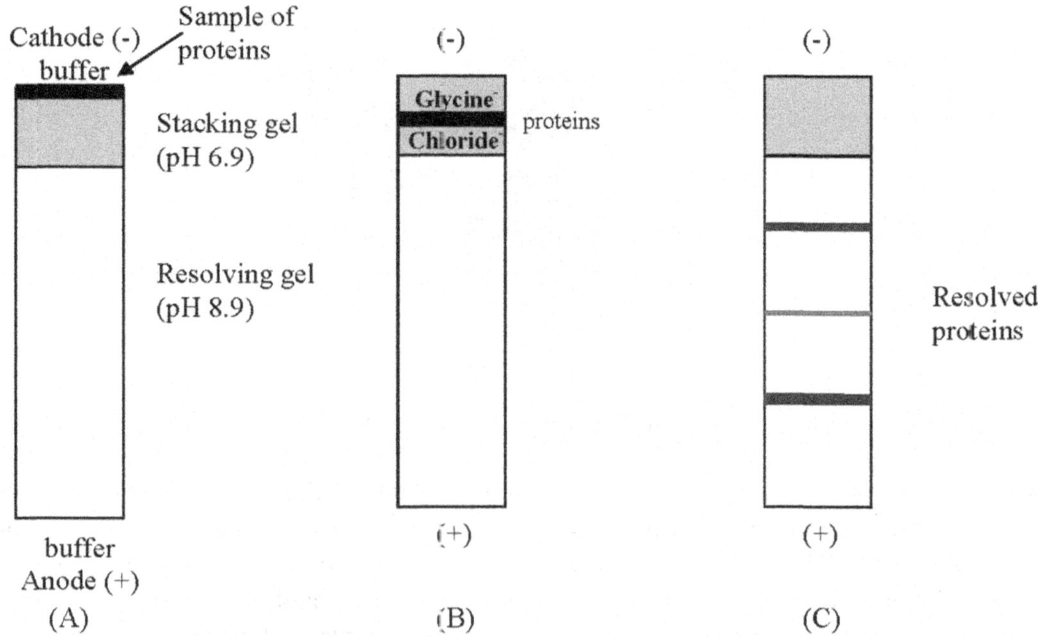


Figure 5.6 – Movement of proteins through discontinuous gel electrophoresis: A) before electrophoresis; B) movement of proteins through the stacking gel influenced by the movement of Cl⁻ and glycine⁻; C) resolving of proteins in the resolving gel. (See text for details.)

## 5.3 SODIUM DODECYL SULFATE POLYACRYLAMIDE GEL ELECTROPHORESIS (SDS-PAGE)

Sodium dodecyl sulfate polyacrylamide gel electrophoresis (SDS-PAGE) is used to determine the molecular weight of a protein, protein sample purity, or the number and size of protein subunits. Discontinuous polyacrylamide gel electrophoresis is carried out in the presence of the detergent sodium dodecyl sulfate, $CH_3$-$(CH_2)_{11}$-$OSO_3^-$ $Na^+$, and a thiol, such as ß-mercaptoethanol.

The ß-mercaptoethanol reduces the disulfide linkages in the protein, creating either linear protein molecules or disrupting interactions between subunits. The result is a mixture of linearized protein chains. The sodium dodecyl sulfate molecules also denature the proteins and bind to the protein in a ratio of 1.4 g SDS to 1 g of protein, which is typically one molecule of SDS for every two amino acids. This denaturation and binding covers the protein uniformly with a negative charge and creates a protein with the same shape (linear, random coil). The bound SDS masks the native charge of the protein and creates linear molecules in which the charge is proportional to the length. Recall that the velocity of a particle is equal to $\frac{Eq}{f}$ , the applied voltage (E), charge ($q$), and frictional coefficient ($f$). The frictional coefficient is dependent on size and shape of the molecule. If the shape is the same and the charge is proportional to the length for proteins bound to SDS, the only difference between the proteins would be the length of the chain. The proteins can then be electrophoretically separated based on molecular weight. Figure 5.7 shows an example of an SDS-PAGE gel.

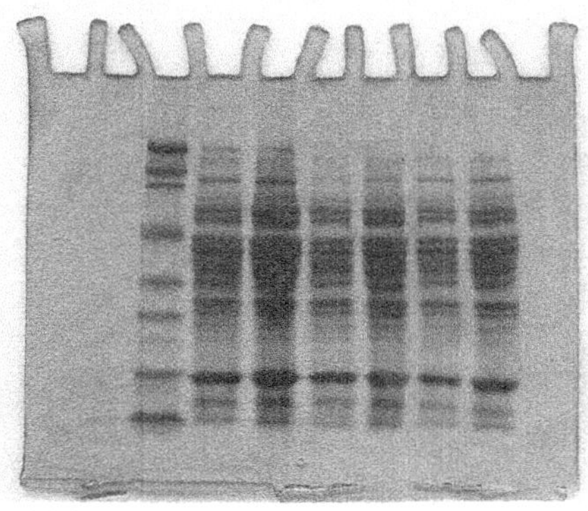

Figure 5.7 – A sample SDS-PAGE gel stained with Coomassie Blue.

Since proteins are separated by molecular weight in SDS-PAGE, this method can be used to determine the molecular weight of an unknown protein. A mixture of standard proteins is separated using SDS-PAGE. The relative mobility of each protein is determined as the distance traveled by the protein band divided by the distance traveled by the tracking dye. (For details on observing bands in gels and tracking dyes, see Sections 5.4 and 5.5). A plot of log of molecular weight versus relative mobility of protein band (Figure 5.8) should be linear. The standard curve can be used to interpolate the molecular weight of an unknown protein. For example, if an unknown protein was run on the same gel as the molecular weight markers and had a relative migration of 0.5, the interpolated log of the molecular weight for this protein from the standard curve would be 4.79. The molecular weight could then be determined by raising ten to 4.79, which gives an approximated molecular weight for the protein of 62,000 Da. A new standard curve must be created for each gel.

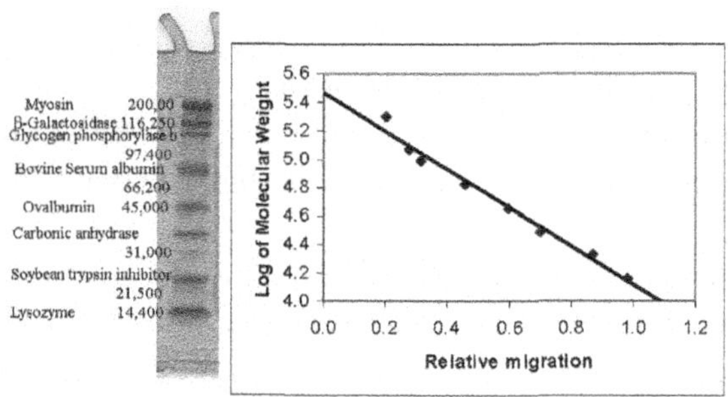


Figure 5.8 – Standard curve for molecular weight determination of proteins derived from the molecular weight standards run on an SDS-PAGE, shown on the left of the plot. The molecular weights of the proteins are provided in Daltons.

Care should be taken when preparing polyacrylamide gels, as acrylamide is a potent neurotoxin. Gloves should be worn at all times when handling acrylamide. Precast gels are now available and are frequently used in laboratories to avoid the handling of the toxic solutions required for polymerization of the gels.

Samples for SDS-PAGE are prepared by mixing the protein sample (maximum protein per lane of the gel should not exceed 20-40 µg total protein) with a loading buffer that contains a tracking dye (see Section 5.4), glycerol, SDS, and ß-mercaptoethanol. After mixing, the sample should be boiled for 2-10 minutes. The heating ensures complete denaturation of the proteins. **Note** – if the solution turns yellow during heating, the sample is too acidic. Sodium hydroxide should be added to the sample until it returns to blue. If the sample is loaded under acidic conditions, the proteins may not migrate correctly, resulting in smearing or abnormal band formation. Following heating, the sample should be centrifuged briefly (for one to several seconds) in a microcentrifuge. This will remove any precipitated protein or debris that may cause streaking on the gel. The samples should be loaded into the wells of the gel using a micropipetter or syringe.

## 5.4 TRACKING DYES

The progress of electrophoresis is often monitored using tracking dyes, which are added to the sample prior to loading into the wells. The dyes are small, negatively charged molecules that travel through the gels faster than most proteins and nucleic acids. Common tracking dyes include Bromophenol Blue and Xylene Cyanol. Table 5.4 shows the size of a DNA fragment that comigrates with the tracking dyes in different concentrations of acrylamide gels.

Table 5.4 – Size of DNA (bp) that Migrates at the Same Rate as a Tracking Dye.

| Acrylamide (% w/v) (29:1 molar ratio of acrylamide to bis-acrylamide) | Bromophenol Blue | Xylene Cyanol |
|---|---|---|
| 3.5 | 100 | 460 |
| 5.0 | 65 | 260 |
| 8.0 | 45 | 160 |
| 12.0 | 20 | 70 |
| 15.0 | 15 | 60 |
| 20.0 | 12 | 45 |

## 5.5 STAINING OF GELS – DETECTIONS OF MACROMOLECULES IN GELS

Unless a protein is colored, after electrophoresis, the gel will contain no bands except for the tracking dye. In order to visualize the location of the biomolecules, a stain must be applied to the gel. Protein gels are stained in a number of different manners, and it should be noted that different proteins may interact differently with a given stain. Uniform staining of all proteins will not be achieved with any currently available stain.

Coomassie Brillant Blue dye is the most common stain and can be used to bind to proteins in gels and stain them blue. (See Chapter 3 for the structure of the dye and its use in spectrophotometric detection of proteins.) After the gel is destained, blue bands can be seen. (See Figure 5.8.) Coomassie dye is typically dissolved in an aqueous solution of methanol and acetic acid (0.25% Coomassie in 5:5:1 $H_2O$, methanol, acetic acid). The gel is soaked for a brief period of time in the dye solution. The gel is then destained by soaking in the same solvent mixture. The detection limit for Coomassie Blue is 0.1–1 μg of protein per band.

Silver staining was first introduced in 1979 as a more sensitive protein stain. It also reveals other macromolecules, including nucleic acids and polysaccharides. The basic mechanism of silver staining involves the reduction of the silver ions to metallic silver by formaldehyde at the protein bands. The detection limit for silver staining of proteins in a gel is 2–10 ng of protein per band.

Fluorescent stains, such as SYPRO Ruby or SYPRO Orange, have equivalent sensitivity as silver stains (1–10 ng of protein per band). These stains are specific for proteins and exhibit no staining of nucleic acids. SYPRO Ruby was specifically formulated for staining 2-D polyacrylamide gels but can be used to stain other polyacrylamide gels. Both of these stains can be visualized using a standard UV transilluminator and gel documentation system.

Nucleic acids are frequently stained using ethidium bromide (see Chapter 3). A gel containing nucleic acids is soaked in a solution of ethidium bromide and washed briefly in water or a buffer. The DNA bands, which appear as red-orange bands, are visualized using ultraviolet light. Double-stranded DNA binds to ethidium bromide with a higher affinity than single-stranded DNA or RNA, and ethidium bromide exhibits greater fluorescence when bound to DNA as opposed to the free dye. This technique works well to visualize double-stranded DNA. An alternative to staining the gel with ethidium bromide is to incorporate the dye into the gel matrix. One thing to keep in mind is that the presence of the dye in the gel support retards the movement of the nucleic acids in the gel by 10–15%.

Nucleic acids can also be stained in gels using methylene blue, silver salts, and fluorescent dyes other than ethidium bromide. Methylene blue can be used to stain DNA in agarose gels or RNA that has been transferred to nitrocellulose filters. The gel or filter is soaked in 0.001–0.0025% methylene blue in an appropriate buffered solution and then visualized on a visible light box. Silver staining of nucleic acids is similar to that described for silver staining of proteins. Fluorescent dye staining of nucleic acids in gels is similar to that described for ethidium bromide.

An alternative method of visualization of nucleic acids in gels is to label the gels with radioactive isotopes or a chemiluminescent label. Nucleic acids can be labeled with $^{32}$P on the 3' or 5' end of the molecule. The radioactive isotope can be detected by autoradiography. The gel containing the radioisotope is exposed to a piece of film, and the radioactive material exposes the film at the site of the band. After development of the film, the bands on the gel can be visualized.

Chemiluminescent probes have been developed to avoid the environmental and safety issues associated with handling radioactive materials. Nucleic acids are labeled with fluorescent tags or are cross-linked to reporter molecules, such as Biotin. The reporter molecule binds to avidin and streptavidin that are labeled with a fluorophore. The fluorescent tags can be detected, and the labeled molecules can be visualized. Proteins can also be detected using chemiluminescent probes.

## 5.6 ISOELECTRIC FOCUSING AND TWO-DIMENSIONAL ELECTROPHORESIS OF PROTEINS

Isoelectric focusing (IEF) is a form of electrophoretic separation that separates molecules as a function of pH. Recall from Chapter 2 that proteins are charged species and that the charge on the molecule depends on the pH of the solution. At a pH equal to the isoelectric point (pI) of a protein, the charge on the protein is zero, and the protein will not migrate in the presence of an electric field. If the pH is less than the pI, the protein is positively charged and will migrate toward the negatively charged electrode. If the pH is greater than the pI, the protein will be negatively charged and will migrate toward the positively charged electrode.

In IEF, a pH gradient is constructed within the matrix material. The protein is applied to the material, and a voltage is applied. A molecule will move in the presence of the current toward the electrode of opposite charge. During the migration, the protein moves through a pH gradient, and as a result of changes in pH, the protein will either gain or lose protons. (See Chapter 2 for more details about amino acid/protein titrations.) As the protein changes its ionization state, its charge will change, and its mobility will decrease. The movement will continue until it reaches a pH at which its charge is zero. At this point, the molecule will remain present at that zone of the gel. Figure 5.9 shows an illustration of the separation of proteins by IEF.

The pH gradient is constructed with the acidic pH closest to the anode (positive electrode) and the basic pH closest to the cathode. The gradients can be broad, such as pH 2-8, or narrow, such as pH 7-8. The gradient is frequently formed using carrier ampholytes, a mixture of polyanionic and polycationic molecules. When placed in an electric field, the ampholytes will arrange in the gel matrix and create a pH gradient based on their intrinsic charges. The protein sample is then introduced to the gel, and the current is applied. The protein will move through the gradient until it reaches its pI. This enables proteins to be separated based on their charge.

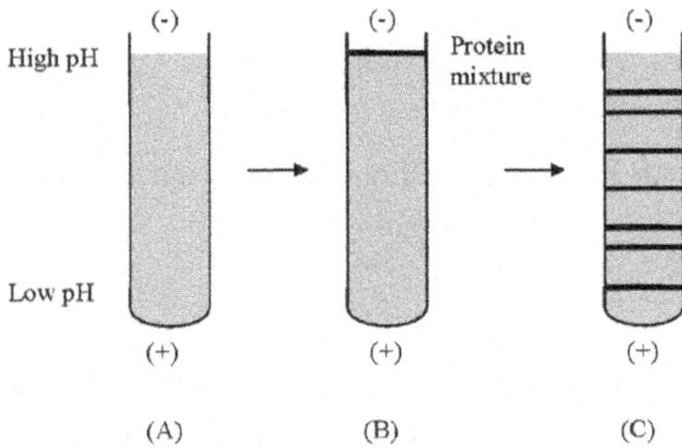



Figure 5.9 – Illustration of protein separation through isoelectric focusing: A) a pH gradient is established in the gel matrix by incorporating ampholytes into the gel and applying an electric current; B) the protein sample is introduced to the gel, and an electric field is applied; C) the proteins separate according to their pIs. The bands closest to the anode (+) have a more acidic pI, whereas those closest to the cathode (-) have a more basic pI.

Two-dimensional electrophoresis is a separation technique that enables the resolution of proteins by both isoelectric point and by molecular weight in a single gel. The first-dimensional separation of the proteins occurs through IEF. The proteins are typically separated in an IEF gel in a glass tube. After the proteins are resolved through a pH gradient, the IEF gel is placed on an SDS-PAGE gel, and the proteins are separated in the second dimension by molecular weight. More than 1,000 proteins can be resolved using two-dimensional electrophoresis.

## 5.7 IMMUNOCHEMICAL METHODS

Antibodies are proteins that are produced by the immune system that bind to foreign substances. The substances that cause the immune system to produce antibodies are immunogens or antigens. Antigens are large-molecular-weight proteins or polysaccharides that are frequently found on the surfaces of viruses and cells. A portion of the antigen called an epitope binds to receptors on immune cells as well as antibodies.

Antibodies are also called immunoglobulins, and they consist of four polypeptides, two heavy chains and two light chains, connected in a Y-shaped molecule, as shown in Figure 5.10. The antigen-binding sites occur in the arms of the Y, and the antigen is recognized by both the heavy and light chains at that site. The region of the antibody structure that differs between types of immunoglobin occurs at and near the antigen-binding sites. These variable domains dictate the specificity of individual antibodies. Antibodies are divided into five major classes, IgM, IgG, IgA, IgD, and IgE, based on their structure and immune function.

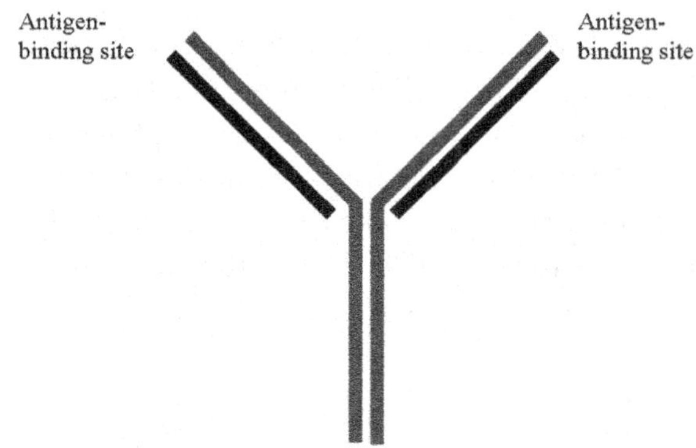


Figure 5.10 – Basic structure of an immunoglobulin. The heavy chains are shown in red, and the light chains are shown in blue.

Antibodies are very specific molecules and can be used as reagents in the identification and analysis of antigens (proteins or polysaccharides). Immunochemical methods or assays use labeled antibodies or antigens to detect antibodies or antigens in samples. There are generally numerous immunochemical techniques used, including radioimmunoassay, enzyme-linked immunoassays, and immunoblotting. The first two techniques will be discussed in this section. Immunoblotting will be discussed in Section 5.8.

Radioimmunoassay (RIA) was first developed in 1960 by S. A. Berson and R. S. Yalow as a means to determine the concentration of insulin in blood plasma. The technique involves a radioactive antigen and an antibody specific for that antigen. Figure 5.11 provides a pictorial representation of the basic procedure of RIA. A mixture of the radioactive antigen and antibody is created. If this mixture is exposed to some unlabelled antigen, a competition for the binding sites between the labeled and unlabeled antigen occurs. A standard curve of the ratio of concentration of bound radioactive antigen to unbound radioactive antigen versus the concentration of unlabeled antigen is created. The amount of antigen in an unknown sample is determined by interpolation on the standard curve.

Enzyme-linked immunoabsorbent assay (ELISA) is a variation of RIA that does not involve radioisotopes. An antigen or a mixture of antigens is applied to the surface of a container, typically a 96-well plate. This sample could be from blood plasma. Any unoccupied sites on the plate are blocked with a nonspecific protein, such as casein from milk. A sample containing antibodies specific for the antigen is applied to the well. The antibody will only bind to sites containing the specific antigen. The solution is then removed, and the well is washed. After washing, a secondary antibody, which is specific for the first antibody, is applied. This antibody will only bind to those sites containing the primary antibody. The secondary antibody contains a linked enzyme that will catalyze a colorimetric reaction. After application of the secondary antibody, the plate is washed to remove unbound antibody, and then the substrate for the enzymatic reaction is applied. The enzyme will convert the substrate to a colored product, and the intensity of the color produced is related directly to the amount of protein of interest in the sample. Figure 5.12 briefly illustrates the steps involved in ELISA.

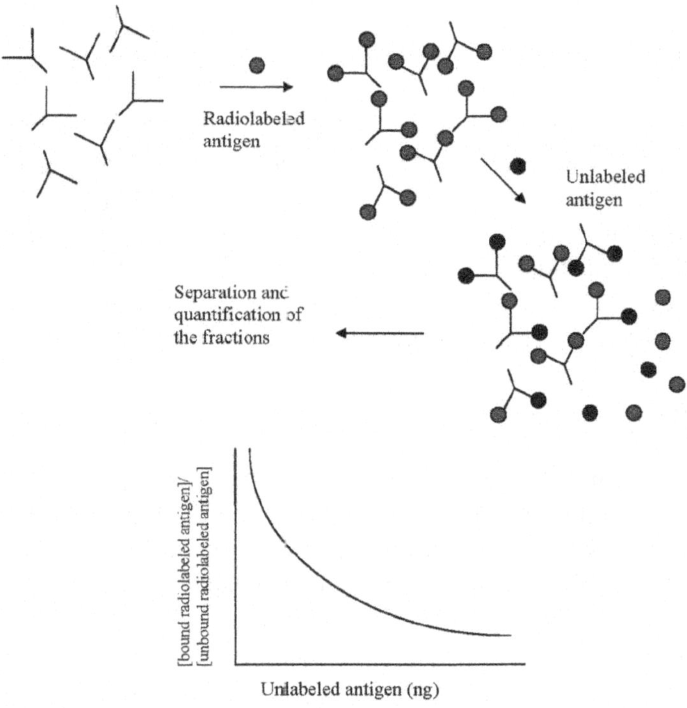


Figure 5.11 – Radioimmunassay (RIA). For details about the assay, refer to the text.

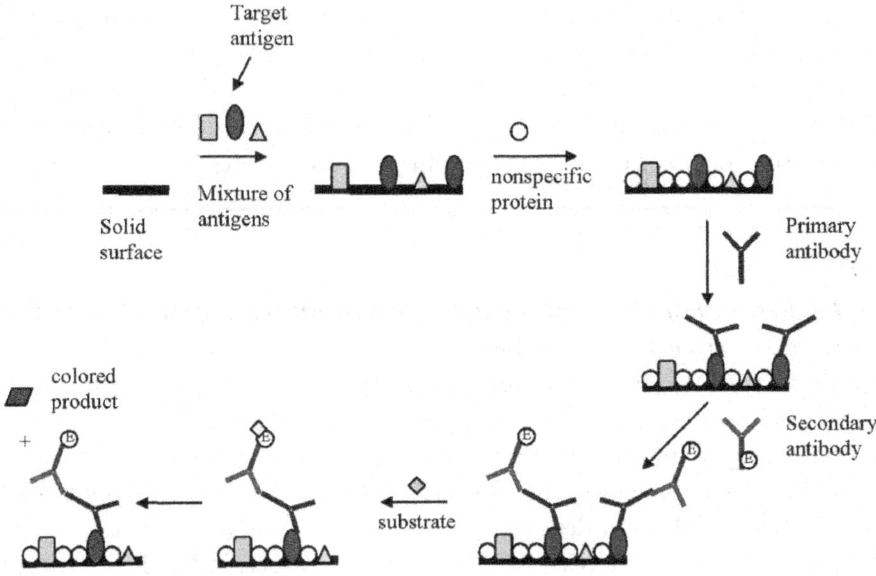


Figure 5.12 – Enzyme-linked Immunosorbent Assay (ELISA). (See text for details.)

# 5.8 PROTEIN AND NUCLEIC ACID BLOTTING

The staining and detection methods described in Section 5.5 reveal the presence of protein or nucleic acids in gels. They often do not provide information on the presence of specific characteristics or sequences. Blotting techniques have been developed in order to determine the location of discrete macromolecules. In this technique, the samples are transferred from the gel onto a piece of nitrocellulose filter or other specialized paper that binds the macromolecules but excludes the matrix materials. The filter paper is then treated with a specific probe that detects the molecule of interest.

### Southern Blotting

The first blotting technique was developed in 1975 by J. Southern. A mixture of double-stranded DNA fragments is separated by electrophoresis. The separated DNA fragments are transferred by blotting to a nitrocellulose or nylon filter. The filter with the samples from the gel is exposed to buffered solution containing a labeled DNA probe that is complementary to a DNA sequence of interest. If the DNA imbedded in the paper has a complementary sequence, it will bind to the probe through DNA-DNA hybridization. The paper is then exposed to film, or the probe is detected in another fashion in order to discern the location of the DNA segment of interest.

### Northern Blotting

An adaptation of the Southern blot is used to detect specific RNA sequences and is called Northern blotting. RNA molecules (usually mRNA) are separated using gel electrophoresis and the RNA is transferred to a filter. The filter is incubated with a probe, usually a labeled single-stranded DNA molecule. The location of the probe is determined by audoradiography or chemiluminescence.

### Western Blotting

Western blotting is a blotting technique in which proteins from a gel are detected by a staining method called immunoblotting. The overall process of immunoblotting is very similar to that described for ELISA in Section 5.7. The proteins are transferred from a gel to a nitrocellulose membrane. The membrane is then blocked with a non-specific protein such as casein or bovine serum album. This blocking prevents the non-specific binding of the antibody directly to the nitrocellulose. The filter/membrane is then treated with a specific primary antibody. After washing, a secondary antibody that is conjugated to an enzyme is applied. The protein bands of interest (those with the primary and secondary antibodies attached) are then detected through the application of a substrate that binds to the enzyme and is converted to a colored complex.

## Science in Action

### Box 5.1 – Restriction Fragment Length Polymorphisms and DNA Fingerprinting

Different individuals exhibit slight but unique sequence differences in DNA fragments that are produced by digestion with restriction enzymes. (Restriction enzymes will be discussed in detail in Chapter 8.) These variations in the DNA are called restriction fragment length polymorphisms (RFLPs) and are used as markers, or DNA fingerprints. RFLP has been used to confirm the identities of individuals from bones or other biological samples uncovered at murder sites. These fragments have also been used to determine paternity or maternity of individuals. The detection of specific RFLPs to help pinpoint individual identification utilizes the Southern blot.

In conclusion, electrophoresis is a powerful tool for separation and identification of biological molecules. It can be used to determine the purity of a sample, the presence of a specific macromolecule, or the size of a biological molecule. The different types of electrophoretic techniques have been briefly described in this chapter. For further information, see the references listed below.

## REFERENCES AND FURTHER READING

Bollag, D. M., Rozycki, M. D., and S. J. Edelstein. *Protein Methods*. 2nd Edition. New York, NY: Wiley and Sons, Inc., 1996.

Boyer, R. *Modern Experimental Biochemistry*. 3rd Edition. San Francisco, CA: Benjamin Cummings, 2000.

Gasteiger E., et al. ExPASy: The Proteomics Server for In-depth Protein Knowledge and Analysis. *Nucleic Acids Res.,* 31, 2003: 3784-3788.

Nelson, D. L., and M. M. Cox. *Lehninger Principles of Biochemistry*. 4th Edition. New York, NY: W.H. Freeman and Co, 2005.

Sanbrook, J., and D. W. Russell. *Molecular Cloning. A Laboratory Manual.* 3rd Edition. Cold Spring Harbor, NY: Cold Spring Harbor Laboratory Press, 2001.

Walker, J. M. *The Protein Protocols Handbook*. Totowa, NJ: Humana Press, 2002.

# Experiment 13. Native Gel Separation of Proteins

## Purpose of the Experiment

Nondenaturing gel electrophoresis (native gel electrophoresis) separates proteins based on their native size and charge. Most proteins are negatively charged at pH 8.8, so the gel separations will be carried out in buffered solutions with a high pH. The purpose of this experiment is to examine the native separations of mixtures of proteins using nondenaturing gel electrophoresis. Your instructor may suggest that you compare your results from this experiment to those in Experiment 14, in which the same mixtures are separated under denaturing conditions. Depending on equipment availability or time constraints, Experiments 13, 14, and/or 15 could be run simultaneously.

## Prelaboratory Assignment

Using appropriate references, determine the approximate charge, shape, and size of the proteins used in the separation. You will use cytochrome c, myoglobin, hemoglobin, and bovine serum album in both experiments. Predict the order of separation on the native gel of the proteins based on your findings.

## Materials

Precast Tris-HCl Polyacrylamide Gels, 12% separating gel, 4% stacking gel
If precast gels are not available, students must cast their own using
       30% acrylamide, 0.8% bis-acrylamide solution
       Separating gel buffer – 1.5 M Tris-Cl (pH 8.8)
       Stacking gel buffer – 0.5 M Tris (pH 6.8)
       10% ammonium persulfate
       TEMED
       Glass plates, spacers, casting apparatus
5X Electrophoresis buffer (15.1 g Tris, 72.0 glycine, water to 1L; do not adjust pH) For
       this experiment, you will need to dilute to 1X buffer.
Sample buffer (7 mL 0.5 M Tris-Cl (pH 6.8), 3.0 mL glycerol, 1.2 mg bromophenol blue,
       water to 10 mL)
Protein marker mixture containing at least 6 different proteins (obtain information about
       the composition from your instructor or the manufacturer)
Protein solutions of 3 or 4 known proteins from table below (1 mg/mL) as directed by
       your instructor. If you are running this experiment in conjuction with Experiment
       14, use the same proteins that you used or will use in that experiment.

| Protein | Molecular Weight (Da) |
|---|---|
| Cytochrome c (horse) | 12,400 |
| α-lactalbumin (bovine) | 14,200 |
| Lysozyme (egg white) | 14,600 |
| Myoglobin (horse) | 17,600 |
| ß-lactoglobulin (bovine) | 18,400 |
| Trypsinogen (bovine) | 24,000 |
| Albumin (egg white) | 44,300 |
| Hemoglobin (bovine) | 64,500 |
| Bovine serum albumin | 66,000 |
| Glycogen phosphorylase b | 97,400 |
| ß-galactosidase | 116,250 |
| Myosin | 200,00 |

Unknown protein sample (1 mg/mL solution)

Electrophoresis apparatus

Staining solution (1.0 g Coomassie Blue R-250, 450 mL methanol, 100 mL glacial acetic
acid, water to 1 L)

Destaining solution (50 mL methanol, 70 mL glacial acetic acid, water to 1 L)


## Safety

Wear goggles. It is recommended that precast gels be obtained commercially to avoid exposure to
unpolymerized acrylamide. Acrylamide, unpolymerized, is a potential neurotoxin and skin irritant. Acrylamide should be handled with gloves and in a hood as much as possible. Dispose of the
solutions at the end of the experiment in the proper manner, as indicated by the instructor.


## Experimental Methods

If precast gels are available, skip to the section on *Prepare protein sample and load gel*. These
instructions are written in general terms; consult your instructor or manufacturer's instructions
about use of specific electrophoresis apparatuses.

*Pour Separating Gel*

These recipes are for a gel of dimensions 0.75 mm x 7 cm x 8 cm. Prepare 5 mL of separating gel
and 2.5 mL of stacking gel even though less will actually be used in the gel. If necessary, you will
need to adjust the recipes to suit your gel size.

1) Assemble the gel plates according to the manufacturer's instructions. Make sure this step
   is completed properly to ensure that the gels do not leak.

2) Put the comb into the gel sandwich, and mark a line on the plate 2 mm below the bottom of the comb.

3) Prepare the separating gel by mixing together the following items in a 25 mL sidearm vacuum flask:

> 2.0 mL of 30% acrylamide, 0.8% bis-acrylamide solution
>
> 1.25 mL of separating gel buffer
>
> 1.75 mL of water

Stopper the sidearm flask, and connect the arm to a vacuum. Degas the sample under vacuum for 10 to 15 minutes. Add 25 µL of 10% ammonium persulfate and 5 µL of TEMED to the solution.

4) Transfer the gel solution to the glass plate sandwich by adding the solution with a pipet down the side of one of the spacers. Fill the plates up to the level of the mark you made in Step 2.

5) Carefully pipet a layer of water-saturated isobutyl alcohol on the surface of the gel solution. Allow gel to stand at room temperature to polymerize for 30-60 minutes.

## *Pour Stacking Gel*

1) Remove the isobutyl alcohol from the gel, and rinse with a small amount of electrophoresis buffer.

2) Prepare the stacking gel by mixing together the following items in a 25 mL sidearm vacuum flask:

> 0.4 mL of 30% acrylamide, 0.8% bis-acrylamide solution
>
> 0.75 mL of stacking gel buffer
>
> 1.85 mL of water

Stopper the sidearm flask, and connect the arm to a vacuum. Degas the sample under vacuum for 10 to 15 minutes. Add 20 µL of 10% ammonium persulfate and 5 µL of TEMED to the solution.

3) Transfer the gel solution to the glass plate sandwich by adding the solution with a pipet down the side of one of the spacers. Fill the plates up to a level approximately 1 cm from the top of the plates.

4) Carefully insert the comb into the solution. Avoid trapping air bubbles under the comb. Pipet additional stacking gel around the comb, if necessary.

5) Allow gel to stand at room temperature to polymerize for 30-60 minutes.

## *Prepare Protein Samples and Load Gel*

1) Remove the comb carefully from the gel and carefully rinse the wells with some electrophoresis buffer.

2) Insert the gel into the electrophoresis chamber, as instructed by the manufacturer. Fill the reservoirs of the electrophoresis chamber with electrophoresis buffer. Check the system for leaks.

3) Combine protein samples with sample buffer. Dilute the protein samples 1:1 with the sample buffer in a microcentrifuge tube. The volume of the sample will depend on the size of the comb used in the polymerization. For example, if the well holds 10 µL, prepare a 15 µL sample, and then 10 µL will be loaded into the gel.

4) Spin the samples in a microcentrifuge for 1 second to remove any debris.

5) Load the samples into the wells of the gel using a Hamilton syringe or micropipetter.

   Load the samples in this order: (This is assuming a 10-well gel is used; adjust as necessary. Any lanes without a protein sample should contain sample buffer.)
   Lane 1: sample buffer only
   Lane 2: protein marker mixture
   Lane 3: known protein sample #1
   Lane 4: known protein sample #2
   Lane 5: known protein sample #3
   Lane 6: known protein sample #4
   Lane 7: unknown sample
   Lane 8: markers
   Lanes 9 and 10: sample buffer only

6) Connect the power supply to the electrophoresis apparatus. Turn the power supply on to 100–200 V (constant current).

7) Allow the electrophoresis to continue until the tracking dye migrates to approximately 1 cm from the bottom of the gel.

8) Turn off the power supply and disassemble the apparatus.

### *Staining of the Gel*

1) Wearing gloves, carefully disassemble the gel sandwich and transfer the gel to a small container. Make sure you place the gel in the container so you know which side is left.

2) Overlay the gel with the staining solution. Allow the gel to stain for at least 20 minutes. The gel can be soaked in the staining solution for longer than 20 minutes or overnight, if necessary.

3) Carefully pour out the staining solution. Avoid flipping or folding the gel.

4) Overlay the gel with destaining solution. Allow the gel to destain completely by changing the destaining solution occasionally and by allowing the gel to remain in the solution overnight.

5) Visualize the bands on the gel and photograph the gel. Alternatively, the gel can be dried, then documented.

6) Calculate the relative mobilities for the bands on your gel. Construct a standard curve of log of molecular weight versus relative mobility for the proteins in the molecular marker mixture. Determine the molecular weights of your known and unknown proteins using the standard curve.

## Questions

1) Did the separation occur as you predicted? If not, explain the differences between the expected and observed results.

2) Normally, SDS-PAGE gels are used to determine the molecular weights of proteins using the standard curve of log of molecular weight versus relative mobility. In this experiment, you created such a standard curve. Is the curve useful? Should native gel electrophoresis be used to determine the molecular weights of proteins?

3) If you conducted Experiment 14, or plan to conduct Experiment 14 in the near future, comment on the differences and similarities of your results. Does one type of gel electrophoresis provide information that differs from the other type? Under what situations would non-denaturing conditions be favored over denaturing conditions?

4) Based on your findings, what is the identity of your unknown sample? Explain.

# Experiment 14. Separation of Proteins Using SDS-PAGE

## Purpose of the Experiment

Sodium dodecyl sulfate-polyacrylamide gel electrophoresis (SDS-PAGE) is a common and rapid method for characterizing, comparing, and analyzing protein samples. This technique can also be used to determine the molecular weights of proteins. The purpose of this experiment is to examine the separations of mixtures of proteins using SDS-PAGE. Your instructor may suggest that you compare your results from this experiment to those in Experiment 13, in which the same mixtures are separated under non-denaturing conditions. Depending on equipment availability or time constraints, Experiments 13, 14, and/or 15 could be run simultaneously.

## Prelaboratory Assignment

In SDS-PAGE, the SDS and 2-mercaptoethanol denature the protein, and the SDS binds uniformly on the surface of the protein, creating linear molecules with charges that are proportional to their length. Using appropriate references, determine the approximate charge, shape, and size of the proteins used in the separation. Predict the order of separation of the proteins by SDS-PAGE.

## Materials

Precast Tris-HCl Polyacrylamide Gels, 12% separating gel, 4% stacking gel
If precast gels are not available, students must cast their own using
      30% acrylamide, 0.8% bis-acrylamide solution
      Separating gel buffer – 1.5 M Tris-Cl (pH 8.8), 0.4% SDS
      Stacking gel buffer – 0.5 M Tris (pH 6.8), 0.4% SDS
      10% ammonium persulfate
      TEMED
      Glass plates, spacers, casting apparatus
5X SDS-PAGE electrophoresis buffer (15.1 g Tris, 72.0 glycine, 5 g SDS, water to 1L;
      do not adjust pH) For experiment, dilute to 1X.
SDS-PAGE sample buffer (7 mL 0.5 M Tris-Cl (pH 6.8), 3.0 mL glycerol, 1 g SDS,
      0.93 g DTT or 0.2 mL of 2-mercaptoethanol, 1.2 mg bromophenol blue, water to
      10 mL)
Protein marker mixture containing at least 6 different proteins (obtain information about
      the composition from your instructor or the manufacturer)
Protein solutions of 3 or 4 known proteins from table below (1 mg/mL) as directed by
      your instructor. If you are running this experiment in conjuction with Experiment
      13, use the same proteins that you used in that experiment.

| Protein | Molecular Weight (Da) |
|---|---|
| Cytochrome c (horse) | 12,400 |
| α-lactalbumin (bovine) | 14,200 |
| Lysozyme (egg white) | 14,600 |
| Myoglobin (horse) | 17,600 |
| ß-lactoglobulin (bovine) | 18,400 |
| Trypsinogen (bovine) | 24,000 |
| Albumin (egg white) | 44,300 |
| Hemoglobin (bovine) | 64,500 |
| Bovine serum albumin | 66,000 |
| Glycogen phosphorylase b | 97,400 |
| ß-galactosidase | 116,250 |
| Myosin | 200,00 |

Unknown protein sample (1 mg/mL solution)
Electrophoresis apparatus
Staining solution (1.0 g Coomassie Blue R-250, 450 mL methanol, 100 mL glacial acetic
	acid, water to 1 L)
Destaining solution (50 mL methanol, 70 mL glacial acetic acid, water to 1 L)


## Safety

Wear goggles. It is recommended that precast gels be obtained commercially to avoid exposure to unpolymerized acrylamide. Acrylamide, unpolymerized, is a potential neurotoxin and skin irritant. Acrylamide should be handled with gloves and in a hood as much as possible. Dispose of the solutions at the end of the experiment in the proper manner, as indicated by the instructor.


## Experimental Methods

If precast gels are available, skip to the section on *Prepare protein sample and load gel*. These instructions are written in general terms; consult your instructor or manufacturer's instructions about use of specific electrophoresis apparati.

### *Pour the Separating Gel*

These recipes are for a gel of dimensions 0.75 mm x 7 cm x 8 cm. Prepare 5 mL of separating gel and 2.5 mL of stacking gel, even though less will actually be used in the gel. If necessary, you will need to adjust the recipes to suit your gel size.

1) Assemble the gel plates according to the manufacturer's instructions. Make sure this step is completed properly to ensure that the gels do not leak.

2) Put the comb into the gel sandwich, and mark a line on the plate 2 mm below the bottom of the comb.

3) Prepare the separating gel by mixing together the following items in a 25 mL sidearm vacuum flask:

> 2.0 mL of 30% acrylamide, 0.8% bis-acrylamide solution
>
> 1.25 mL of separating gel buffer
>
> 1.75 mL of water

Stopper the sidearm flask, and connect the arm to a vacuum. Degas the sample under vacuum for 10 to 15 minutes. Add 25 µL of 10% ammonium persulfate and 5 µL of TEMED to the solution.

4) Transfer the gel solution to the glass plate sandwich by adding the solution with a pipet down the side of one of the spacers. Fill the plates up to the level of the mark you made in Step 2.

5) Carefully pipet a layer of water-saturated isobutyl alcohol on the surface of the gel solution. Allow gel to stand at room temperature to polymerize for 30-60 minutes.

### Pour Stacking Gel

1) Remove the isobutyl alcohol from the gel and rinse with a small amount of electrophoresis buffer.

2) Prepare the stacking gel by mixing together the following items in a 25 mL sidearm Vacuum flask:

> 0.4 mL of 30% acrylamide, 0.8% bis-acrylamide solution
>
> 0.75 mL of stacking gel buffer
>
> 1.85 mL of water

Stopper the sidearm flask, and connect the arm to a vacuum. Degas the sample under vacuum for 10 to 15 minutes. Add 20 µL of 10% ammonium persulfate and 5 µL of TEMED to the solution.

3) Transfer the gel solution to the glass plate sandwich by adding the solution with a pipet down the side of one of the spacers. Fill the plates up to a level approximately 1 cm from the top of the plates.

4) Carefully insert the comb into the solution. Avoid trapping air bubbles under the comb. Pipet additional stacking gel around the comb, if necessary.

5) Allow gel to stand at room temperature to polymerize for 30-60 minutes.

### Prepare Protein Samples and Load Gel

1) Remove the comb carefully from the gel and carefully rinse the wells with some electrophoresis buffer.

2) Insert into the electrophoresis chamber as instructed by the manufacturer. Fill the reservoirs of the electrophoresis chamber with electrophoresis buffer. Check the system for leaks.

3) Combine protein samples with sample buffer. Dilute the protein samples 1:1 with the sample buffer. The volume of the sample will depend on the size of the comb used in the polymerization. For example, if the well holds 10 µL, prepare a 15 µL sample, and 10 µL will be loaded into the gel.

4)  Boil each sample for 5-10 minutes. Spin the samples in a microcentrifuge for 1 second to remove any debris.

5)  Load the samples into the wells of the gel using a Hamilton syringe or micropipetter.

Load the samples in this order: (This is assuming a 10-well gel is used; adjust as necessary. Any lanes without a protein sample should contain sample buffer.)
Lane 1: sample buffer only
Lane 2: protein marker mixture
Lane 3: known protein sample #1
Lane 4: known protein sample #2
Lane 5: known protein sample #3
Lane 6: known protein sample #4
Lane 7: unknown sample
Lane 8: markers
Lanes 9 and 10: sample buffer only

6)  Connect the power supply to the electrophoresis apparatus. Turn the power supply on to 100–200 V (constant current).

7)  Allow the electrophoresis to continue until the tracking dye migrates to approximately 1 cm from the bottom of the gel.

8)  Turn off the power supply and disassemble the apparatus.

## *Staining of the Gel*

1)  Wearing gloves, carefully disassemble the gel sandwich and transfer the gel to a small container. Make sure you place the gel in the container so you know which side is left.

2)  Overlay the gel with the staining solution. Allow the gel to stain for at least 20 minutes. The gel can be soaked in the staining solution for longer than 20 minutes or overnight, if necessary.

3)  Carefully pour out the staining solution. Avoid flipping or folding the gel.

4)  Overlay the gel with destaining solution. Allow the gel to destain completely by changing the destaining solution occasionally and by allowing the gel to remain in the solution overnight.

5)  Visualize the bands on the gel and photograph the gel. Alternatively, the gel can be dried, then documented.

6)  Calculate the relative mobilities for the bands on your gel. Construct a standard curve of log of molecular weight versus relative mobility for the proteins in the molecular marker mixture. Determine the molecular weights of your known and unknown proteins using the standard curve.

## Questions

1) Did the separation occur as you predicted? If not, explain the differences between the expected and observed results.

2) Normally, SDS-PAGE gels are used to determine the molecular weights of proteins using the standard curve of log of molecular weight versus relative mobility. In this experiment, you created such a standard curve. Is the curve useful? How does this curve compare to the one created for native electrophoresis in Experiment 13 (if you completed this experiment). Which form of gel electrophoresis should be used to determine the molecular weights of proteins?

3) If you conducted Experiment 13, or plan to conduct Experiment 13 in the near future, comment on the differences and similarities of your results. Does one type of gel electrophoresis provide information that differs from the other type? Under what situations would non-denaturing conditions be favored over denaturing conditions?

4) Based on your findings, what is the identity of your unknown sample? Explain.

# Experiment 15. Identification of Proteins by Western Blotting

## Purpose of the Experiment

Western blotting is an important technique for identification and characterization of proteins. In this experiment, proteins in bovine serum will be separated by electrophoresis and transferred onto nitrocellulose paper. Proteins interacting with anti-bovine IgG will be detected by Western blotting. Depending on equipment availability or time constraints, Experiments 13, 14, and/or 15 could be run simultaneously.

## Prelaboratory Assignment

Answer the following questions:

1) What is the role of each of the following in the procedure:
   a) nitrocellulose
   b) anti-Bovine IgG

2) Using appropriate references, determine the chemical reaction catalyzed by alkaline phosphatase. How does this reaction relate to the detection of proteins using Western blotting?

## Materials

Precast Tris-HCl Polyacrylamide Gels, 12% separating gel, 4% stacking gel
If precast gels are not available, students must cast their own using the procedures listed
    in Experiments 13 and 14
5X SDS-PAGE electrophoresis buffer (15.1 g Tris, 72.0 glycine, 5 g SDS, water to 1L;
    do not adjust pH)
Native electrophoresis buffer (15.1 g Tris, 72.0 glycine, water to 1L; do not
    adjust pH)
SDS-PAGE sample buffer (7 mL 0.5 M Tris-Cl (pH 6.8), 3.0 mL glycerol, 1 g SDS,
    0.93 g DTT or 0.2 mL of 2-mercaptoethanol, 1.2 mg bromophenol blue, water to
    10 mL)
Native electrophoresis sample buffer (7 mL 0.5 M Tris-Cl (pH 6.8), 3.0 mL glycerol,
    1.2 mg bromophenol blue, water to 10 mL)
Protein marker mixture containing at least 6 different proteins (obtain information about
    the composition from your instructor or the manufacturer) – prestained
IgG (1 mg/mL)
Bovine serum
Electrophoresis apparatus
Staining solution (1.0 g Coomassie Blue R-250, 450 mL methanol, 100 mL glacial acetic
    acid, water to 1 L)
Destaining solution (100 mL methanol, 100 mL glacial acetic acid, water to 1 L)
Nitrocellulose filter
Transfer buffer (1.93 g Tris (pH 8.1-8.4), 9 g glycine, water to 1 L)

Whatman 3 MM filter paper
Tris-buffered saline (TBS) (10 mM Tris-HCl (ph 7.5), 150 mM NaCl)
3% BSA in TBS or 1% nonfat dry milk in TBS
0.5% BSA in TBS
Anti-Bovine IgG – Alkaline Phosphatase Antibody produced in rabbit
Alkaline phosphatase buffer (0.1 M Tris-HCl (pH 9.5), 0.1 M NaCl, 5 mM $MgCl_2$)
BCIP solution (50 mg/mL 5-bromo-4-chloro-3-indolyl phosphate in 100%
        dimethylformamide)
NBT solution (50 mg/mL $p$-nitro blue tetrazolium chloride in 70% dimethylformamide)
20 mM EDTA in TBS


## Safety

Wear goggles. It is recommended that precast gels be obtained commercially to avoid exposure to unpolymerized acrylamide. Acrylamide, unpolymerized, is a potential neurotoxin and skin irritant. Acrylamide should be handled with gloves and in a hood as much as possible. Dispose of the solutions at the end of the experiment in the proper manner, as indicated by the instructor.


## Experimental Methods

Please refer to Experiments 13 and 14 to prepare native and SDS-PAGE gels. The samples will be prepared in duplicate once for each of the electrophoresis techniques (i.e., separation under both non-denaturing and denaturing conditions).

### *Prepare Protein Samples and Load Gel*

See note above. This procedure should be conducted twice—using both native gel electrophoresis and SDS-PAGE.

1) Remove the comb carefully from the appropriate gel and carefully rinse the wells with some electrophoresis buffer (make sure you use the correct buffer for the type of electrophoresis to be conducted).

2) Insert into the electrophoresis chamber as instructed by the manufacturer. Fill the reservoirs of the electrophoresis chamber with electrophoresis buffer. Check the system for leaks.

3) Combine samples (marker mixture, IgG, and bovine serum) with sample buffer (make sure you use the correct sample buffer). Dilute the protein samples 1:1 with the sample buffer. The volume of the sample will depend on the size of the comb used in the polymerization. For example, if the well holds 10 µL, prepare a 15 µL sample, and then load 10 µL into the gel.

4) If conducting SDS-PAGE, boil each sample for 5-10 minutes. If not, skip this step.

5) Spin the samples in a microcentrifuge for 1 second to remove any debris.

6) Load the samples into the wells of the gel using a Hamilton syringe or micropipetter.

   Load the samples in this order: (This is assuming a 10-well gel is used; adjust as necessary. Any lanes without a protein sample should contain sample buffer.)
   Lane 1: sample buffer only
   Lane 2: protein marker mixture
   Lane 3: IgG sample
   Lane 4: bovine serum
   Lane 5: sample buffer
   Lane 6: sample buffer
   Lane 7: protein marker mixture
   Lane 8: IgG sample
   Lane 9: bovine serum
   Lane 10: sample buffer only

7) Connect the power supply to the electrophoresis apparatus. Turn the power supply on to 100–200 V (constant current).

8) Allow the electrophoresis to continue until the tracking dye migrates to approximately 1 cm from the bottom of the gel.

9) Turn off the power supply, and disassemble the apparatus.

10) Remove the gel from the apparatus and carefully cut the gel between lanes 5 and 6. Half of the gel will be used for the immunotransfer and blot. (Continue with the sections on *Transfer of Proteins* and *Development of the Immunoblot.*) The other half of the gel will be stained with Coomasie Blue. (Continue with *Staining of the Gel.*)

## *Transfer of Proteins*

These instructions are written in general terms; consult your instructor or manufacturer's instructions about use of specific transfer apparati.

1) Soak a piece of nitrocellulose membrane that is approximately the same size as your gel in transfer buffer for 15-20 minutes at room temperature. Always handle the nitrocellulose with gloved hands or tweezers.

2) Using the manufacturer's instructions, assemble a polyacrylamide gel-membrane sandwich. The typical order of assembly is
   a) a piece of Whatman 3 MM paper (pre-wetted with the transfer buffer)
   b) the gel – arrange the gel on the filter paper so that it is flat and all bubbles are removed. To avoid transfer of contaminants from the skin, wear gloves and use tweezers as much as possible.
   c) wet piece of nitrocellulose membrane
   d) wetted piece of Whatman 3 MM paper

3) Roll over the surface of the sandwich gently with a small test tube to remove any air bubbles.

4) Place the sandwich into the transfer apparatus, following the manufacturer's instructions.

5) Transfer the proteins by applying 100 V constant voltage to the sample for 1 hour or as specified by your instructor or the manufacturer.

### Immunodetection

1) Disconnect and disassemble the apparatus.

2) Remove the nitrocellulose filter using tweezers or wearing gloves. Place the filter in a small container or plastic bag.

3) Remove the gel from the sandwich and stain it to ensure complete transfer of the proteins. (See *Staining of Gel* section.)

4) Cover the membrane in 3% BSA in TBS and incubate for 30 minutes to 1 hour. This will prevent non-specific antibody binding to the membrane. If necessary, the membrane can be stored overnight or for several days in 3% BSA in TBS with 1 mM $NaN_3$ added.

5) Pour off the BSA solution and wash the nitrocellulose three times in TBS.

6) Soak the nitrocellulose in 0.5% BSA/TBS containing the anti-IgG antibody in a ratio of 1:30,000 for at least an hour or up to overnight.

7) Pour off the antibody solution and rinse the nitrocellulose three times in TBS for 10 minutes each.

### Development of the Immunoblot

1) Wash the nitrocellulose membrane in alkaline phosphatase buffer for 5 minutes.

2) Prepare the developing reagent by mixing 66 μL of NBT solution with 10 mL of alkaline phosphatase buffer. After mixing well, add 33 μL of BCIP solution. Use this solution within 1 hour of mixing.

3) Remove the nitrocellulose from the alkaline phosphatase buffer. Add 10 mL of the solution from Step 2 to the nitrocellulose filter.

4) Incubate at room temperature or 37°C (the higher temperature speeds the reaction) until the bands appear. This generally occurs within 30 minutes, but overnight incubation is sometimes needed.

5) The reaction can be stopped by rinsing the filter in 20 mM EDTA in TBS.

### Staining of the Gel

1) The unblotted gel and the gel remaining after blotting should be stained for evidence of protein. Wearing gloves, transfer the gel to a small container. Make sure you place the gel in the container so you know which side is left.

2) Overlay the gel with the staining solution. Allow the gel to stain for at least 20 minutes. The gel can be soaked in the staining solution for longer than 20 minutes or overnight, if necessary.

3) Carefully pour out the staining solution. Avoid flipping or folding the gel.

4) Overlay the gel with destaining solution. Allow the gel to destain completely by changing the destaining solution occasionally and by allowing the gel to remain in the solution overnight.

5) Visualize the bands on the gel and photograph the gel. Alternatively, the gel can be dried then documented.

## Questions

1) Did the proteins completely transfer from the gels to the nitrocellulose? If not, what experimental changes should be made to ensure complete transfer?

2) Did you observe any differences in the migration of the IgG on the different types of gels (denaturing versus non-denaturing)? Does one type of gel electrophoresis provide information that differs from the other type? Under what situations would non-denaturing conditions be favored over denaturing conditions?

3) Did the antibody recognize the proteins in the same manner on either type of gel? Explain.

4) Most Western blot procedures call for incubation with the primary antibody and then incubation with a secondary antibody. (See Sections 5.7 and 5.8 for details.) This procedure only involved one antibody. Explain why.

5) Is there any IgG in the bovine serum? Did any other proteins react with the antibody? Explain your findings.

# 6

# Sample Preparation and Partial Purification of Biological Molecules

Prior to some experiments, you must prepare the sample and/or isolate materials from cellular materials or complex mixtures. The purification of specific macromolecules will depend on the properties and characteristics of the molecules. Even though the methods will differ, some similarities exist among the methods. First, a cellular or tissue sample must be disrupted or lysed in order to create an extract from which the macromolecule will be purified. Some of the most common means for cell or tissue disruption include homogenization, sonication, vortexing, and chemical lysis. The cellular extract is then used directly or fractionated to partially purify the material. Initial fractionation or partial purification can be obtained through centrifugation, filtration, dialysis, precipitation, and/or extraction. This chapter will discuss all of these techniques in detail. Further purification or separation can be achieved with chromatography and/or electrophoresis, which are described in Chapters 7 and 5, respectively.

## 6.1 PREVENTING/LIMITING CONTAMINATION OF SOLUTIONS

When conducting studies with cells or enzymes that may be affected by contaminants (i.e., chemicals such as metal ions and/or microbes such as bacteria or yeast), it is important that a sterile environment be maintained. Most of the solutions created in a biochemical laboratory contain water. The quality of the water is therefore important. Tap water contains ions, trace organics, and microbes and should not be used routinely to prepare solutions. Water is purified most frequently through distillation or deionization. Distilled water contains low quantities of ions and nonvolatile substances but may contain volatile substances, such as carbon dioxide. Deionized water is created by passing water through mixed ion-exchange resins (for more details on ion-exchange chromatography, see Chapter 7). Deionized water is low in ionic compounds but often contains low quantities of ultraviolet-absorbing organic materials that are washed from the resins. These materials may interfere with spectroscopic measurements. (See Chapter 3.) Many laboratories contain ultrapurification water systems that pass deionized water through additional ion exchange resins, polishing systems, and filters to remove additional contaminants from the water. Ultrapure water tends to have the least amount of interferences for future analytical techniques.

In order to prevent microbial growth, all items used to prepare the solutions should be sterile. The glassware and equipment used can be sterilized by autoclaving. Many sterilized items can also be purchased commercially. Sterile items should be handled with care to prevent contamination. You should wear gloves and work in a sterile environment such as a laminar flow hood. Solutions can either be sterilized by autoclaving, by filtering through a small-pore filter (diameter of 0.2 μm or less), or by exposure to ultraviolet radiation. In addition, small amounts of sodium azide (0.02% by weight) can be added to solutions to prevent microbial growth. Also, refrigeration of solution will limit growth of microorganisms.

Other contaminants that may impact biochemical experiments include oxidizing or reducing agents, enzymes such as proteases or RNases, and metal ions. If protein oxidation is of concern, reducing agents such as 2-mercaptoethanol or dithiothreitol can be added to a solution used in the purification or assay. In order to prevent protein degradation by proteases, protease inhibitors, such as phenylmethanesulfonyl fluoride (PMSF) and aprotitin, are frequently employed. RNases are hardy and powerful enzymes that can easily ruin an experiment involving RNA. Vigilance in handling materials, using RNase-free equipment, and use of RNase inhibitors, such as diethypyrocarbonate (DEPC), can limit the effects of RNases on experiments. Metal ion contamination is often prevented by using chelators such as ethylene diaminetetraacetic acid (EDTA).

## 6.2 CELL DISRUPTION METHODS

The lysis of cellular samples must occur in such a way that disrupts the cell wall or membrane and creates an extract in which the molecule of interest has not been damaged or destroyed. The method of choice will depend on the specific macromolecule to be isolated, and experimentation may be necessary to maximize purification, ensuring the largest yield and greatest activity of the molecule. Common methods for cell disruption include homogenization, sonication, vortexing, and chemical lysis.

### *Homogenization*

Homogenization is the mechanical disruption of tissues through the use of either a blender or a Teflon pestle in a narrow glass container. Lysis using a homogenizer is rapid (5-10 minutes) and is most commonly used for soft animal tissues and many plant tissues. If necessary, a tissue sample is crudely chopped into smaller pieces (often 1 cm cubes) using a knife. The tissue is mixed with 3–5 volumes of buffer and is placed in the homogenizer. Following the manufacturer's specifications, the sample is homogenized.

### *Disruption of Cell Suspensions*

Cell suspensions are frequently lysed using sonication, grinding, or glass bead vortexing. Sonication is a technique that uses ultrasonic vibrations to disrupt cell walls or membranes. The sample is mixed with at least 2 volumes of buffer and sonicated for several minutes. The degree of sonication should be monitored carefully, and foaming of the solution should be avoided. Excessive sonication can cause protein denaturation and DNA sheering.

Cell suspensions can also be lysed by grinding the sample in a mortar and pestle with an abrasive material, such as sand or alumina. Using a chilled mortar and pestle, a paste of cells (about 20 g) and sand or alumina (approximately 40 g) is ground for 5–10 minutes. Following grinding, the mixture is mixed with additional buffer (about 60 – 100 mL), and the sand or alumina and cellular debris is removed by centrifugation. (See Section 6.3.)

A variation on grinding with an abrasive material is vortexing the sample in the presence of glass beads (500 µm diameter). A pellet of cells is suspended in an equal volume of lysis buffer and placed into an appropriate-sized plastic tube. Small samples can be lysed in eppendorf tubes. Chilled glass beads (1–3 g per gram of cells) are added to the tube. The sample is vortexed for 1 minute, several times. The sample should be chilled on ice for 1 minute between vortexing.

### *Chemical Lysis Methods*

Cells can also be lysed by treatment with enzymes and/or detergents. Yeast and bacteria can be disrupted using enzymes such as lysozyme or zymolyase. Lysozyme and zymolyase are enzymes that degrade the backbones of carbohydrates in the cell walls of bacteria and yeast, respectively. A cellular suspension is mixed with the enzyme (1–10 mg/mL concentration) and incubated at 20–37°C for 10-20 minutes. Cultured animal cells are often disrupted with detergents. Detergents are amphoteric molecules that interact with membrane lipids. A pellet of cells is mixed with a buffer containing detergents such as Triton X-100, Tween 20, or Sodium dodecyl sulfate (SDS). Following gentle mixing or vortexing, the sample is incubated on ice for 10–60 minutes.

## 6.3 CENTRIFUGATION

Frequently following cell lysis, centrifugation is used to separate the cellular debris from substances of interest or fractionate materials in solution. Centrifugation is also used to pellet cells from a cell suspension. Another use of centrifugation is to measure physical properties of some particles. A centrifuge is a device that separates or concentrates substances suspended in a medium based on size, density, shape, viscosity of the medium, and the rotor speed. In biochemistry, most particles separated using centrifugation are cells, cellular organelles, small organisms such as bacteria or viruses, and large macromolecules such as proteins and nucleic acids.

The sedimentation rate of a particle can be described using Stokes' Law:

$$V_g = \frac{d^2(P_p - P_l)g}{18\eta} \quad ,$$

where

$V_g$ = sedimentation velocity of a particle
d = particle diameter
$P_p$ = particle density
$P_l$ = liquid density
g = gravity acceleration
$\eta$ = viscosity of the liquid

Stokes' law states that the sedimentation velocity of a particle is related to the size and density of the particle and the viscosity and density of the liquid. Particles that have densities similar to that of the liquid or are very small in size will sediment at a very low rate. Increasing the viscosity of the medium will decrease the sedimentation rate of the particle. Since the rate of settling is related to physical properties, the sedimentation rate can therefore be used to characterize particles and is frequently employed in biochemistry for this purpose.

Typically, the sample to be "spun" is placed in a centrifuge tube, which is inserted into a rotor. A rotor is constructed of a dense metal that dissipates heat quickly and generates momentum so that once it is spinning, it requires little energy to keep it moving. Centrifuges generally operate under vacuum and are refrigerated to reduce heating generated by frictional forces as the rotor spins. The centrifugal force, which can be thousands of times greater than gravitational force, is used in these instruments to increase the settling rate of the particle. The centrifugal force generated is proportional to the square of the rotational rate of the rotor (in rpm) and the distance between the rotor center and the centrifuge tube.

$$F = m \times \omega^2 \times r, \text{ where}$$
$$F = \text{centrifugal force}$$
$$m = \text{mass of the particle}$$
$$\omega = \text{angular velocity of the rotor (radians/seconds)}$$
$$r = \text{radius (distance of the particle from the axis of rotation)}$$

Centrifugal forces are reported relative to the gravitational force (xg). For example, 15,000 xg would be 15,000 times greater than the standard force of gravity.

Different rotor sizes can be used in the same instrument to achieve different centrifugation conditions. Each centrifuge has a special graph, a nomograph, which relates rotation rate (rpm) to centrifugal force (xg) for each size of rotor it uses. Examples of a nomograph can be found in a centrifuge instruction manual, at some manufacturer's websites, and in laboratory technique manuals. Because rotors and centrifuges come in all shapes and sizes, the universal unit for centrifugation is centrifugal force (xg). Therefore, in order to repeat a literature procedure, you need to use a nomograph to convert the xg to rpm for the rotor and centrifuge you plan to use.

There are four types of centrifuges that you may encounter in a biochemical (or similar) laboratory:

1) Clinical centrifuges – These tabletop instruments can reach speeds of up to 10,000–15,000 revolutions per minute (rpm). These instruments can be used to sediment larger particles such as cells, but cannot separate smaller substances like organelles or macromolecules.

2) Microfuges – These small tabletop instruments can run at speeds up to 14,000 rpm. They use microcentrifuge (eppendorf) tubes that typically hold ≤ 2.0 mL. These instruments can be used to pellet small or large particles from small volumes of solutions.

3) High-speed centrifuges – The rotors in these instruments can reach speeds up to 30,000 rpm. These instruments are used to sediment cellular organelles, as well as most macromolecules and larger particles.

4) Ultracentrifuges – These instruments run at very high speeds up to 100,000 rpm. These instruments can be used to pellet macromolecules such as nucleic acids.

### *Differential Centrifugation*

There are two commonly used types of centrifugation: differential centrifugation (rate-zonal centrifugation) and density gradient centrifugation. Differential centrifugation is the more commonly used form of centrifugation. In differential centrifugation, the particles in solution are separated in a centrifugal field over time, based on differing sedimentation velocities. The larger and more-dense particles will settle to the bottom of the tube as the pellet, and the lighter species, will remain suspended in the supernatant. Figure 6.1 shows a diagram of differential centrifugation applied to cellular tissues. Beginning with a cellular extract, the sample is placed in a centrifuge tube and "spun" at the centrifugal force indicated above the arrow. After the centrifugation, the resulting separation is shown below the tube. The supernatant is transferred to another tube and "spun" at the next centrifugal force indicated. For example, after centrifugation at 1000 xg, the pellet contains cellular debris, nuclei, and whole cells. The supernatant contains other cellular components and is centrifuged at 20,000 xg to pellet out mitochondria, peroxisomes, and lysosomes.

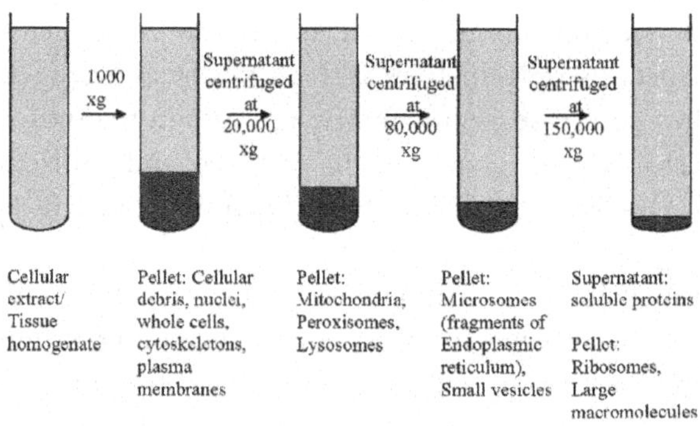


Figure 6.1 – An example of sub-cellular fractionization by differential centrifugation.

## *Density Gradient Centrifugation*

Differential centrifugation separates particles based on size and density, and the resulting fractions will often contain mixtures of substances of differing densities and sizes. In density gradient centrifugation, the sample is loaded on top of a density gradient, a fluid gradient whose density increases from top to bottom. The sample will separate based on density, irrespective of size. Reconsider Stokes' law:

$$V_g = \frac{d^2(P_p - P_l)g}{18\eta}$$

If a particle has the same density ($P_p$) as the liquid medium ($P_l$), the sedimentation velocity will be zero, and the particle will not sediment. If $P_p$ is greater than $P_l$, the sedimentation velocity will be greater than zero, and the particle will sediment. If $P_p$ is less than $P_l$, the sedimentation velocity will be less than zero, and the particle will rise to the top. Therefore, when "spun" in the presence of a density gradient, particles will migrate in the centrifugal field to an equilibrium position that has a similar density and will stop moving. There are two types of density gradient centrifugation: zonal centrifugation, in which the density gradient is prepared prior to centrifugation, and isopynic centrifugation, in which the gradient forms during the centrifugation.

## *Zonal Centrifugation*

In zonal centrifugation, a density gradient is created in a centrifuge tube. The gradient is frequently created using an automatic gradient mixer or by carefully layering solutions of decreasing density (concentration) on top of solutions with higher density. An alternative method would be to underlay a solution of higher density beneath a solution of lower density. The cellular extract or sample is then placed on top of the gradient, and the particles will diffuse through the gradient until the density of the particle equals the density of the liquid. A centrifuge is used to accelerate the process of equilibration.

Gradients are often made by layering different concentrations of solutions of increasing viscosity, such as sucrose, glycerol, or Ficoll$^{TM}$. Sucrose is a disaccharide that can be dissolved in water or buffered solutions to create solutions of different densities. Solutions of up to 60% (w/v) of sucrose that have densities up to 1.2 g/mL can be created. Drawbacks to sucrose include that at concentrations greater than 20% (w/v), the viscosity of the solution increases greatly. Molecular separations are retarded in highly viscous solutions and require longer centrifugation times. Also, sucrose is osmotically active and cannot be used to separate intact cells. Carbohydrates also frequently interfere with spectroscopic analyses such as Lowry assays. (See Chapter 3.) Benefits of using sucrose include low cost, ease of availability, and ease in removal from solutions by dialysis or ultrafiltration. Glycerol density gradients are prepared and utilized in a similar manner as sucrose gradients. Typical density ranges for glycerol gradients are 1.0–1.3%.

Ficoll polymers are formed from the copolymerization of sucrose and epichlorohydrin. These copolymers are commercially available in several types, depending on the molecular weight of the polymer. Ficoll solutions can be prepared at varying concentrations, typically up to 50% (w/v), which corresponds to a density of 1.2 g/mL. The Ficoll solutions have a lower viscosity and osmotic pressure than similar sucrose solutions and are, therefore, able to separate intact cells. One drawback to Ficoll is that it cannot be easily removed from solutions using dialysis; however, it can be removed using ion exchange or gel filtration chromatography.

### *Isopynic Centrifugation*

In this technique, the sample is homogeneously mixed with the gradient medium and centrifuged. In the presence of the centrifugal force, the gradient molecules will sediment out slightly and create a density gradient. The sample molecules will travel to a location in the gradient in which the density of the medium equals their density. Isopynic gradients are frequently created with salt such as cesium chloride or organic molecules such as Nycodenz and Percoll. These substances can also be used in density gradients that are prepared before adding the sample, but are more frequently used in isopynic applications.

Cesium chloride is a salt that yields high-density solutions (up to 1.8 g/mL) when dissolved in water or buffered solutions. The solutions are fairly low in viscosity, especially when compared to sucrose. Gradients of cesium chloride are typically used to isolate nucleic acids. The salt at high concentration causes protein precipitation. (See Section 6.5.) Cesium chloride solutions have a high osmotic pressure and cannot be used to isolate intact cells. Following centrifugation, the salt can be easily removed from DNA solutions using dialysis or ion-exchange chromatography.

Percoll is a colloidal silica particle that is coated with polyvinylpyrrolidone. Solutions of varying densities, up to 1.13 g/mL, exhibit low viscosity, and low osmolality; therefore, Percoll can be used to separate intact cells as well as cellular components. Samples can be recovered from Percoll by washing and further centrifugation or by gel filtration or ion-exchange chromatography.

Nycodenz (also called HistoDenz$^{TM}$) is another gradient medium that exhibits high density (up to 1.4 g/mL), low osmolality, low viscosity and low toxicity. It can be used to separate macromolecules as well as intact cells. The structure of Nycodenz is shown in Figure 6.2. The molecule is soluble in water as well as organic solvents such as formamide and dimethylformamide. It can be readily removed from samples by centrifugation, dialysis, ultrafiltration, or chromatography.

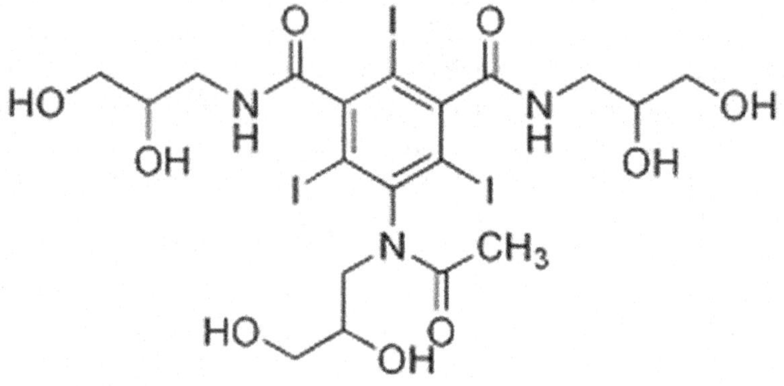

Figure 6.2 – Structure of Nycodenz.

### *Centrifugation Precautions*

Make sure that you have received proper instruction in the use of the centrifuge or have consulted the manual thoroughly. If improperly used, rotors can be ejected at high speeds from a centrifuge, causing injuries to the operator. Make sure that the rotor being used is properly installed on the spindle and that you have selected the appropriate centrifuge tubes. Balance the tubes within the rotor: each tube must be positioned such that a tube of equal mass is present on the other side of the rotor. Proper balancing ensures that the rotor will spin correctly. Do not override any safety features of the centrifuge. Do not leave the centrifuge unattended until it safely reaches maximum speed. If you require assistance with a centrifuge, please consult your instructor.

## 6.4 CONCENTRATION OR FURTHER PREPARATION OF SAMPLES

Frequently, the methods for cellular lysis and extraction result in dilute concentrations of the cellular extract. In order to more properly isolate the compound of interest, the sample frequently needs to be concentrated. The concentration of samples can be accomplished through several methods, including freeze-drying, dialysis, ultrafiltration, and precipitation. Often, after cellular lysis or centrifugation, the sample is in a buffer solution that is incompatible with future analysis. For example, a high salt concentration will cause aberrant bands in electrophoresis. (See chapter 5.) Buffer exchange can be achieved by dialysis and ultrafiltration.

### *Freeze-Drying or Lyophillization*

Freeze-drying is the removal of water from a sample at low temperatures. The water is removed through the process of sublimation, in which ice is directly converted to water vapor. The major use for freeze-drying is the long-term storage of samples. When the majority of the water is removed from the sample, microbial growth and aqueous side reactions that could occur are minimized. One of the downsides to freeze-drying is that the structure of some proteins could be altered, which may destroy enzymatic activity.

### *Dialysis*

Dialysis is one of the oldest techniques used in biochemistry. This technique involves the diffusion of solutes between solutions of different concentrations through a semipermeable membrane. Dialysis can be used for desalting a solution, buffer exchange, or concentration of a solution.

Consider a bag composed of dialysis membrane, shown in Figure 6.3, which contains a solution (Solution A) of a mixture of solutes. The small, circular particles can pass through the membrane, but the dark, oval-shaped solute will be retained. The dialysis bag is placed in a solution (Solution B) that either lacks the solute or contains a lower concentration of the solute than Solution A. The solute will diffuse from Solution A to B at a rate that depends on the concentration difference in diffusible solute between Solutions A and B. The diffusion will continue until the concentration of the solute that can diffuse in Solution A and B is equal, as shown in Figure 6.3 (b). If the purpose of dialysis is to remove the circular particle from Solution A, Solution B should be frequently changed to ensure a concentration difference between A and B exists. This will allow the majority of the particles to be removed, but you can never completely remove all of a substance using dialysis.

The rate of diffusion is also dependent on the surface area of the dialysis membrane. The larger the surface area of the membrane, the greater the rate of diffusion of the solute through the membrane. Another variable that affects the rate of diffusion is the volume of Solution A. If a large volume of Solution A is present in the dialysis container, the solute would have to travel a larger distance before reaching the surface of the membrane where diffusion can occur.

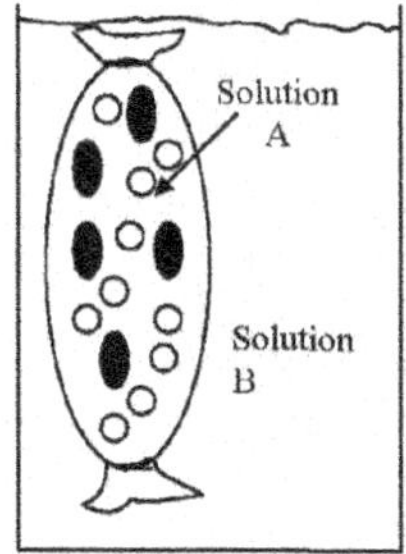

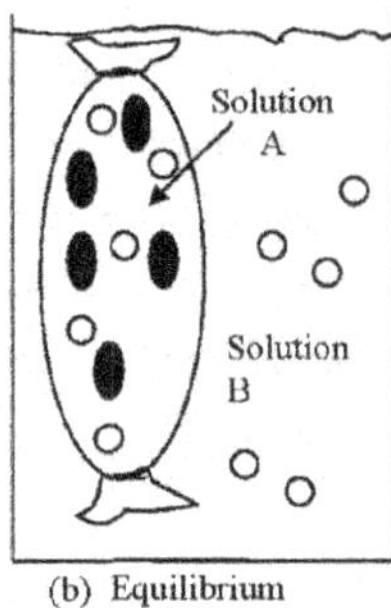


Figure 6.3 - Example of a dialysis system. A solution of two solutes (Solution A) is placed in a dialysis bag and immersed in Solution B. The open-circled solute can pass through the membrane, but the dark oval-shaped molecule will be retained. Based on the concentration difference, the circular solute will travel from Solution A to Solution B.

The sample to be dialyzed is typically placed in a tubular piece of dialysis membrane. The tubing is commercially available in membranes that have different pore sizes between the cellulose fibers. The difference in pore sizes allows scientists to control the type of particles that can pass through the membrane. Manufacturers sell tubing of different molecular weight cutoffs (MWCO). The commonly available MWCOs vary between 100–300,000 Da. The MWCO is defined as the molecular weight at which 90% of the solute will be retained by the membrane. If the tubing has a MWCO of 100,000 Da, particles of less than 100,000 Da would be able to pass through the membrane, particles of greater than 100,000 Da would be retained, and 90% of those particles with an MW of 100,000 Da would be retained. For most biochemical purposes, the MWCO should be selected as half the molecular weight of the solute of interest to be retained.

Dialysis tubing is often sold dried and must be soaked or washed according to the manufacturer's instructions prior to use. After soaking, the tubing is either tied, knotted, or clamped on one end and filled with the solution of interest. The tube should not be filled completely because during dialysis, a small amount of water will travel from Solution B into Solution A by osmosis. If the bag is filled completely, there is a risk that the tubing could burst due to an increase in osmotic pressure.

Typically, dialysis is used to desalt a solution, exchange buffers, or separate molecules. If Solution A contains a higher concentration of salt than Solution B, the salt ions travel from A to B, and after dialysis, the concentration of salt in Solution A would be lower. In a similar manner, buffer molecules can travel from A to B and also from B to A, if an appropriate concentration difference exists. This results in the exchange of the buffer solution between dialysis compartments. Depending on the pore size of the membrane, certain molecules can be filtered out or removed from the sample. In this manner, some size selection can occur through dialysis.

Another use of dialysis is the concentration of a solution. As mentioned before, during dialysis, osmosis occurs and, as a result, Solution A becomes more dilute. If the dialysis tube is placed in a solution of higher osmotic pressure, however, the water would flow from Solution A to Solution B, based on osmosis. This would result in reducing the volume of Solution A and, therefore, concentrating the solution. Dialysis bags are often placed in solutions of sucrose or polyethylene glycol in order to concentrate Solution A. Alternatively, the dialysis tube could be directly placed in granular sucrose, and the water leaves the dialysis bag by osmosis.

### *Ultrafiltration*

Ultrafiltration is a technique that is related to dialysis. It is used to desalt solutions, exchange buffers, or concentrate solutions. When compared with dialysis, ultrafiltration is more rapid but more expensive.

There are multiple types of ultrafiltration devices commercially available. This chapter will describe two that are often encountered by undergraduates. One ultrafiltration device (see Figure 6.4) has a semipermeable membrane (with an appropriate MWCO) near the bottom of the device. Nitrogen gas is passed into a tube at the top of the device. The pressure from the gas forces the solvent and particles that are small enough to pass through the pores in the membrane to exit the chamber and out a tube at the bottom of the chamber. In this manner, the solution volume is reduced, and the sample is concentrated. Also, salt, small contaminants, or buffer components will exit with the solvent, and desalting can be accomplished.

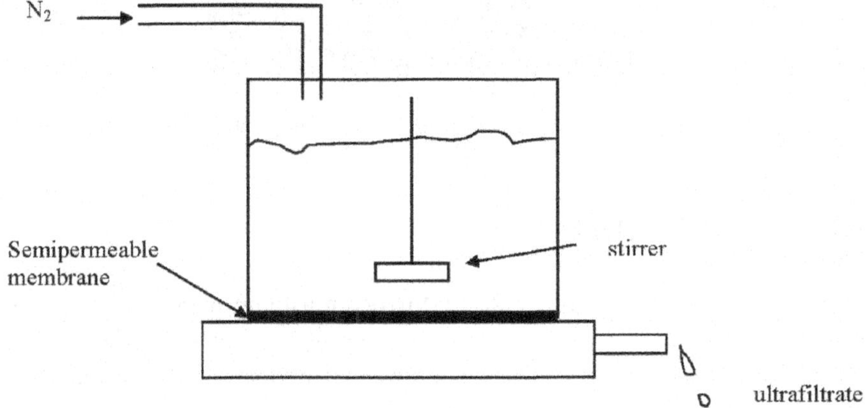


Figure 6.4 – Ultrafiltration device. The solution to be concentrated is placed in a chamber. Nitrogen gas is passed into the chamber, and the solvent and small molecules or ions pass through the membrane and out of the tube at the bottom. (See text for more details.)

As the solvent leaves the chamber, larger molecules, such as biological macromolecules, can build up on the surface of the membrane. These molecules cannot enter the pores, but will often line the surface and block the pores. This layering will limit the flow of the solvent from the chamber and slow the rate of filtration. In order to prevent the slowing of rate and the formation of this layer, the solution inside the chamber is stirred with a magnetic stir bar suspended in the solution.

Ultrafiltration can be used for buffer exchange. A sample is placed in the device and ultrafiltration is carried out until the volume of the solution is reduced. The concentrated sample is dissolved in a new buffer, and this new solution is concentrated through ultrafiltration. The resulting concentrated sample is dissolved again in the new buffer, and ultrafilration is conducted again. This "washing" is repeated several times to accomplish the buffer exchange.

Another type of ultrafiltration device passes the sample through the membrane by centrifugation. Figure 6.5 shows examples of several centrifugal ultrafiltration devices commercially available from Millipore. These devices consist of two parts: a sample reservoir, which contains a membrane with a specific MWCO at the bottom of the chamber, and a collection reservoir. The sample reservoir containing the sample to be desalted or concentrated is placed on top of the collection reservoir. The unit is centrifuged as directed by the manufacturer. During centrifugation, the solvent and molecules with molecular weights smaller than the MWCO will pass from the sample reservoir through the membrane and into the collection reservoir. Depending on the purpose of the separation, the filtrate (the sample passed through the membrane), or the retenate (the concentrated sample in the sample reservoir) is saved and used in future studies.

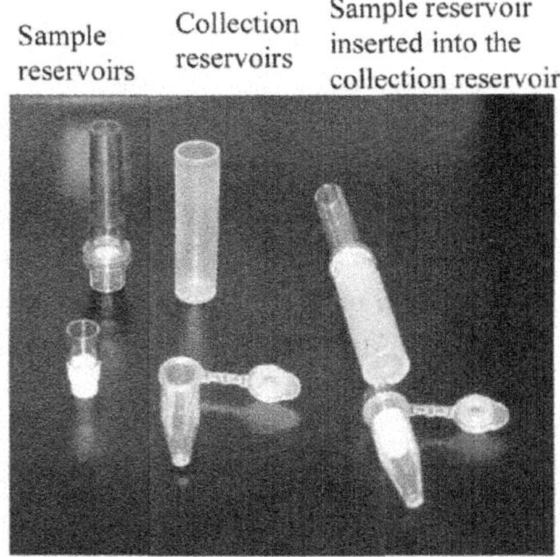


Figure 6.5 –Centrifugal filter device from Millipore.

## 6.5 PRECIPITATION METHODS

Another means of concentrating a sample or further purifying a sample is to precipitate the macromolecule of interest out of solution. A biological molecule can be made less soluble through treatment with a salt, organic solvent, acid, or some combination of these substances. The macromolecule becomes denatured, less soluble, and can be recovered through centrifugation. The pellet will contain the denatured macromolecule.

### *Salting Out of Proteins*

Precipitation of proteins in the presence of salts is one of the classic methods in protein biochemistry. This method is not highly specific and typically does not result in a pure fraction of protein, but this is an inexpensive and fairly rapid means of recovering proteins from solution. In the presence of high concentrations of salt, particularly ammonium sulfate, proteins will aggregate and precipitate out of solution. This is commonly referred to as "salting out."

The solubility of proteins in solution is controlled by various factors, including concentration of salt, pH, and temperature. At low concentrations of salt, the solubility of a protein increases slightly. This process is referred to as "salting in." As discussed in Chapter 2, proteins have intrinsic charges based on the ionization state of the side chains in the structure. Ions of opposite charge in solution will be attracted to these charged groups. This electrostatic interaction stabilizes the charge, attracts the protein into solution, and, therefore, increases solubility.

At a high concentration of salt, the solubility of the protein will decrease, and the protein will precipitate out of solution ("salting out"). When the salt concentration is increased, the salt will interact more with the water, and less water is available to interact with the protein. The protein molecules will begin to aggregate and precipitate from solution.

The amount of precipitation also depends on the type of salt utilized in the experiment. In 1888, Franz Hofmeister studied the effects of salt on the solubility of hen egg white proteins. In general, his findings show the decreasing order of effectiveness of various anions on protein solubility:

Anions: $citrate^{3-} > SO_4^{2-} = tartrate^{2-} > HPO_4^{2-} > CrO_4^{2-} > acetate^- > HCO_3^- > Cl^- > NO_3^- > ClO_3^-$, where $citrate^{3-}$ will be the most effective at precipitating proteins and $ClO_3^-$ least

This series of the relative effectiveness of different ions on protein precipitation is often referred to as the Hofmeister series. A similar series for cations exists, but it is hypothesized that the anions have the greater impact on protein solubility. The exact mechanism through which these ions affect the solubility of proteins is still unclear.

Ammonium sulfate is one of the most common salts employed in protein precipitation. It is classified as one of the salts in the series that will stabilize protein structure during the precipitation. It is also popular because it is highly soluble and relatively inexpensive. Different proteins will precipitate from solution at different concentrations of salt. Most proteins precipitate at 55% ammonium sulfate. The lower concentration end for most proteins is 24% and the upper concentration is 80%. A "salting out" curve can be created for any protein. The curve is a plot of the log of solubility versus the salt concentration. Different proteins will have a distinct linear plot, and the slope of the line is dependent on the protein, the salt involved, the pH, and the temperature.

The basic procedure for ammonium sulfate precipitation of proteins involves adding ammonium sulfate to a solution of protein until a desired concentration is reached. Table 6.1 is a nomogram for determining the amount of ammonium sulfate needed to reach a specified concentration. Consider the following example: the fractionation of proteins in a 50 mL solution will be achieved by adding 25% followed by 50% ammonium sulfate. Initially, the sample contains no ammonium sulfate. In this case, using Table 6.1, one should start in the row with 0 starting percentage and determine the mass of salt needed to create a liter with a 25% final percentage (i.e., 144 g). Therefore, 7.2 g should be added to the 50 mL solution to create a 25% solution. After precipitation and recovery of the insoluble proteins from the 25% solution, the solution contains approximately 25% ammonium sulfate. The volume may increase slightly. In our example, let the volume equal 51 mL. In order to increase the salt concentration to 50%, one uses the table to determine the correct amount. Starting in the row with 25 starting percentage, it can be determined that 157 g is required to convert a liter of the 25% solution to a 50% solution. Therefore, 8.01 g of ammonium sulfate should be added to the 51 mL 25% solution.

After the amount of salt is determined and added, the solution is stirred gently until the salt dissolves. Excessive stirring can cause denaturation of the proteins. Foaming of the solution is an indication of denaturation of proteins. The precipitated proteins are recovered by centrifugation. In general, as the molecular weight of the protein increases, the amount of salt required to precipitate the protein decreases. This method can therefore be used to fractionate a mixture of proteins. The larger proteins will precipitate out first, and the smaller ones will remain in solution. By adding ammonium sulfate in increasing amounts, proteins in solution can be fractionated, and the concentration of salt needed for precipitation can be determined.

### Organic Solvent Precipitation

An organic solvent, such as acetone or ethanol, can be added to a solution to precipitate proteins. The solubility of a protein is dependent on the dielectric constant of the medium in which it is dissolved. Addition of an organic solvent causes a decrease in the dielectric constant and a decrease in the protein solubility. At high concentration, though, organic solvents are often attracted to the hydrophobic portions of protein, which are often in the interior, causing denaturation of the proteins. Therefore, care must be taken to keep the concentration of organic solvents low and to keep the temperature below 10°C to limit denaturation of the proteins. The amount of organic solvent needed to precipitate a protein will depend on the structure of the protein. Typically, the range of organic solvent needed to carry out precipitation is 5–60% (v/v) organic solvent.

The basic procedure for organic solvent precipitation is very similar to that used for salting out of proteins. The organic solvent is added as a percent volume, and it is assumed that the volumes are additive. For example, using ethanol, a 50% (v/v) solution is created by mixing 50 mL of protein solution with 50 mL of ethanol, even though the final volume of the solution will be less than 100 mL. The organic solvent should be chilled before addition to the protein solution, and the two liquids should be mixed together on an ice bath. The resulting solution is stirred gently on ice for 10-15 minutes. The insoluble proteins are collected by centrifugation at 10,000 xg for 10 minutes. The supernatant is decanted, and the pellet is resuspended in cold buffer. If insoluble particles remain in the solution, they are likely denatured proteins and should be removed by centrifugation or filtration.

### Trichloroacetic Acid Precipitation of Proteins

Trichloroacetic acid (TCA) precipitation is another method for recovering proteins from solution. TCA precipitation is frequently used to recover proteins from small volumes of dilute solution prior to SDS-PAGE. A sample containing protein is mixed with an equal volume of 20% TCA (w/v). After vortexing, the sample is incubated on ice or in a freezer for 30 minutes. The protein is then recovered through centrifugation in a microcentrifuge at full speed for 15 minutes. The supernatant is removed, and 300 μL of cold acetone is added to the protein pellet. The solution is centrifuged again for 5 minutes. The supernatant is removed, and the pellet is allowed to dry. The dried pellet is suspended in loading buffer for SDS-PAGE.

Table 6.1 - Mass of Ammonium Sulfate (g) Required to Create One Liter Solutions of Specific Concentrations of Ammonium Sulfate. (ref. Bollag)

| Starting Percentage | Final Percentage | | | | | | | | | | | | | | | | | | | |
|---|---|---|---|---|---|---|---|---|---|---|---|---|---|---|---|---|---|---|---|---|
| | 5 | 10 | 15 | 20 | 25 | 30 | 35 | 40 | 45 | 50 | 55 | 60 | 65 | 70 | 75 | 80 | 85 | 90 | 95 | 100 |
| 0 | 27 | 55 | 84 | 113 | 144 | 176 | 208 | 242 | 277 | 314 | 351 | 390 | 430 | 472 | 516 | 561 | 608 | 657 | 708 | 761 |
| 5 | | 27 | 56 | 85 | 115 | 146 | 179 | 212 | 246 | 282 | 319 | 357 | 397 | 439 | 481 | 526 | 572 | 621 | 671 | 723 |
| 10 | | | 28 | 57 | 86 | 117 | 149 | 182 | 216 | 251 | 287 | 325 | 364 | 405 | 447 | 491 | 537 | 584 | 634 | 685 |
| 15 | | | | 28 | 58 | 88 | 119 | 151 | 185 | 219 | 255 | 292 | 331 | 371 | 413 | 456 | 501 | 548 | 596 | 647 |
| 20 | | | | | 29 | 59 | 89 | 121 | 154 | 188 | 223 | 260 | 298 | 337 | 378 | 421 | 465 | 511 | 559 | 609 |
| 25 | | | | | | 29 | 60 | 91 | 123 | 157 | 191 | 227 | 265 | 304 | 344 | 386 | 429 | 475 | 522 | 571 |
| 30 | | | | | | | 30 | 61 | 92 | 126 | 160 | 195 | 232 | 270 | 309 | 351 | 393 | 438 | 485 | 533 |
| 35 | | | | | | | | 30 | 62 | 94 | 128 | 163 | 199 | 236 | 275 | 316 | 358 | 402 | 447 | 495 |
| 40 | | | | | | | | | 31 | 63 | 96 | 130 | 166 | 202 | 241 | 281 | 322 | 365 | 410 | 457 |
| 45 | | | | | | | | | | 31 | 64 | 97 | 132 | 169 | 206 | 245 | 286 | 329 | 373 | 419 |
| 50 | | | | | | | | | | | 32 | 65 | 99 | 135 | 172 | 210 | 250 | 292 | 335 | 381 |
| 55 | | | | | | | | | | | | 33 | 66 | 101 | 138 | 175 | 215 | 256 | 298 | 343 |
| 60 | | | | | | | | | | | | | 33 | 67 | 103 | 140 | 179 | 219 | 261 | 305 |
| 65 | | | | | | | | | | | | | | 34 | 69 | 105 | 143 | 183 | 224 | 266 |
| 70 | | | | | | | | | | | | | | | 34 | 70 | 107 | 146 | 186 | 228 |
| 75 | | | | | | | | | | | | | | | | 35 | 72 | 110 | 149 | 190 |
| 80 | | | | | | | | | | | | | | | | | 36 | 73 | 112 | 152 |
| 85 | | | | | | | | | | | | | | | | | | 37 | 75 | 114 |
| 90 | | | | | | | | | | | | | | | | | | | 37 | 76 |
| 95 | | | | | | | | | | | | | | | | | | | | 38 |

### Polyethylene Glycol (PEG) Precipitation

Polyethylene glycol is a nonionic water-soluble polymer that precipitates proteins without denaturation. The mechanism of PEG precipitation is still unclear, but it is hypothesized that the PEG interacts with the protein, preventing water solvation from occurring, and the protein solubility is decreased. A sample containing protein is mixed with a PEG solution (often a 50% w/v solution in water) to the desired final concentration of PEG. Most proteins will precipitate with a final PEG concentration of 30% w/v. The sample is stirred for 30–60 minutes. The protein is recovered through centrifugation at 10,000 xg for 10 minutes. The pellet is resuspended in 1–2 volumes of buffer. PEG can be removed from the precipitate using ammonium sulfate precipitation or ion exchange chromatography.

### Ethanol Precipitation of DNA

DNA is often removed from solution using ethanol precipitation. Proteins, salt, and nucleotides remain in solution, so this method provides a means of partially purifying as well as concentrating DNA. The ethanol lowers the dielectric constant of the solution and removes the hydration sphere surrounding the DNA. This exposes the negatively charged backbone of the DNA that binds to the positively charged cations in the solution, typically sodium or ammonium ions. These electrostatic interactions decrease the repulsion between the nucleic acid backbones and facilitates precipitation of the DNA.

The sample is mixed with 0.3 M sodium acetate solution (0.1 volume of 3 M acetate solution, pH 5.2) or with 2.5 M ammonium acetate (0.33 volumes of 7.5 M). After mixing, 2 to 2.5 volumes of 100% ice-cold ethanol is added. For concentrated solutions of DNA (>0.1 µg/mL) the DNA readily precipitates at room temperature and does need to be incubated on ice. Dilute solutions of DNA should be mixed and chilled at -20 to -70°C. The DNA is separated from the solution by centrifugation. The supernatant is removed, and the resulting pellet, which contains the DNA, is washed with 70% ethanol. The tube is centrifuged again, and the supernatant is removed. The pellet is dried and then resuspended in an appropriate buffer.

## 6.6 EXTRACTION

Extraction is another technique that can be used to isolate and possibly purify a substance. Extraction involves shaking or gently mixing a solution containing the substance of interest with an immiscible solvent. Molecules will distribute themselves between the water and the organic solvent in proportion to their relative solubilities in the two solvents. Thus, the extraction can be considered a competition between two immiscible liquids for the solute, with the solute partitioning itself between these two liquids. Upon standing, the solvents form two layers that can be separated. Depending on the nature of the macromolecule, biochemists will use the organic layer or the aqueous layer.

### Phenol Extraction

Phenol extraction is frequently used to remove protein from nucleic acid solutions. The proteins are denatured and are soluble in the phenol, whereas nucleic acids remain in the aqueous layer. A solution containing DNA is mixed with an equal volume of phenol/chloroform/isoamyl alcohol or buffered-saturated phenol. Following vortexing and brief centrifugation, the upper aqueous solution is removed. A small quantity of 8-hydroxyquinoline is frequently added to phenol solutions to remove oxidants that could damage the DNA. The DNA is often recovered after extraction, using ethanol precipitation.

## *Extraction of Lipids*

Lipids, unlike other macromolecules, are more soluble in organic solvents than in water. Organic solvents, such as methanol, chloroform, and ether, are frequently employed in the isolation of lipids from cellular extracts. A commonly used solvent mixture for the extraction is chloroform:methanol:water (1:2:0.8). The cellular extract is mixed with the chloroform mixture, and the layers are allowed to separate. The lipids will enter the chloroform layer, while other cellular components will remain in the methanol water mixture. The chloroform solution can be further separated and analyzed using chromatography.

This chapter introduces some of the more commonly used methods for sample preparation and fractionation. It is not meant to be all-inclusive, but provides a starting point for understanding biochemical sample preparations. Additional information can be obtained from reading the resources cited in References and Further Reading.

# REFERENCES AND FURTHER READING


Bollag, D. M., Rozycki, M. D., Edelstein, S.J. *Protein Methods.* 2nd Edition. New York, NY: Wiley and Sons, Inc., 1996.

Boyer, R. *Modern Experimental Biochemistry.* 3rd Edition. San Francisco, CA: Benjamin Cummings, 2000.

Dennison, C. *A Guide to Protein Isolation.* New York, NY: Kluwer Academic Publishers, 2002.

F. Hofmeister, Zur Lehre von der Wirkung der Salze, *Arch. Exp. Pathol. Pharmakol.* (Leipzig), 24, 1888: 247-260.

Farrell, S. O., and R. T. Ranallo. *Experiments in Biochemistry. A Hands-on Approach.* Florence, KY: Brooks Cole Thomas Learning, 2000.

Herm-Götz, et al. *Toxoplasma gondii* Myosin A and its Light Chain: a Fast, Single-Headed, Plus-End-Directed Motor. *EMBO J.*, 21, 2002: 2149-2158.

Nelson, D. L., and M. M. Cox. *Lehninger Principles of Biochemistry.* 4th Edition. New York, NY: W.H. Freeman and Co., 2005.

Sanbrook, J., and D. W. Russell. *Molecular Cloning. A Laboratory Manual.* 3rd Edition. Cold Spring Harbor, NY: Cold Spring Harbor Laboratory Press, 2001.

Spadaro, A. C. C., et al. Salt Fractionation of Plasma Proteins. *Biochem. Mol. Biol. Educ.*, 31, 2003: 249-252.

Turchi, S. L., and M. Weiss. An Experiment Using Sucrose Density Gradients in the Undergraduate Biochemistry Laboratory. *J. Chem. Educ.*, 65, 1988: 170.

Walker, J. M. *The Protein Protocols Handbook.* Totowa, NJ: Humana Press, 2002.

### *Online References*

Ficoll PM 70, Ficoll PM 400 Data File number 18-1158-27, *Amersham Biosciences.*
    <http://www5.amershambiosciences.com/aptrix/upp01077.nsf/Content/na_home
    page>.

*Millipore Corporation Website.* <http://www.millipore.com/>.

*Spectrum® Laboratories, Inc. Website.* June 2005. <http://www.spectrapor.com/index2.html>.

# Experiment 16. Salting Out of Proteins

## Purpose of the Experiment

Ammonium sulfate can be used to precipitate proteins as a means of concentrating or fractionating proteins. The purpose of this experiment is to examine the effect of ammonium sulfate precipitation on different proteins. Your instructor may ask that you use the same proteins from Experiments 13 and 14 so that comparisons between purification methods can be made.

## Prelaboratory Assignment

1) Refer to Step 6 of the ammonium sulfate precipitation procedure. Assuming that the supernatant that is referenced in that step is 9 mL, calculate the amount of ammonium sulfate that must be added to that volume to create a 40% ammonium sulfate solution. (Refer to Table 6.1 and Section 6.5.)

2) How can the ultrafiltration devices described in Chapter 6 be used to concentrate and desalt the samples?

3) Based on the information in Chapter 6, predict the order of solubility of the proteins in the mixture used in the experiment.

## Materials

Solid ammonium sulfate
A solution containing 3 or 4 known proteins from table below (2 mg/mL of each protein), as directed by your instructor. If you are running this experiment in conjunction with other experiments, use the same proteins in each experiment.

| Protein | Molecular Weight (Da) |
| --- | --- |
| Cytochrome c (horse) | 12,400 |
| α-lactalbumin (bovine) | 14,200 |
| Lysozyme (egg white) | 14,600 |
| Myoglobin (horse) | 17,600 |
| ß-lactoglobulin (bovine) | 18,400 |
| Trypsinogen (bovine) | 24,000 |
| Albumin (egg white) | 44,300 |
| Hemoglobin (bovine) | 64,500 |
| Bovine serum albumin | 66,000 |
| Glycogen phosphorylase b | 97,400 |
| ß-galactosidase | 116,250 |
| Myosin | 200,00 |

Spectrophotometer
Solutions for protein quantification assay (select one from Experiment 5)
Electrophoresis apparatus
Precast Tris-HCl Polyacrylamide Gels, 12% separating gel, 4% stacking gel
If precast gels are not available, students must cast their own using
        30% acrylamide, 0.8% bis-acrylamide solution
        Separating gel buffer – 1.5 M Tris-Cl (pH 8.8)
        Stacking gel buffer – 0.5 M Tris (pH 6.8)
        10% ammonium persulfate
        TEMED
        Glass plates, spacers, casting apparatus
Electrophoresis buffer (15.1 g Tris, 72.0 glycine, water to 1L; do not adjust pH)
Sample buffer (7 mL 0.5 M Tris-Cl (pH 6.8), 3.0 mL glycerol, 1.2 mg bromophenol blue,
      water to 10 mL)
Protein marker mixture containing at least 6 different proteins (obtain information about
      the composition from your instructor or the manufacturer)
Staining solution (1.0 g Coomassie Blue R-250, 450 mL methanol, 100 mL glacial acetic
      acid, water to 1 L)
Destaining solution (100 mL methanol, 100 mL glacial acetic acid, water to 1 L)


## Safety

Wear goggles. Dispose of the solutions at the end of the experiment in the proper manner, as indicated by the instructor.


## Experimental Methods

*Ammonium Sulfate Precipitation of Proteins*

1) Place 10 mL of the protein solution into a centrifuge tube. Chill the tube in an ice bath. Slowly add 1.13 g of ammonium sulfate to the tube. Do not add all of the salt at the same time. Gently shake the tube after each addition until the salt dissolves. Keep the tube on ice during the addition. This should create a solution that contains 20% ammonium sulfate. (Refer to Table 6.1 and Section 6.5.)

2) Incubate the tube on ice for 5 minutes after all of the salt has dissolved.

3) Centrifuge the tube at 10,000 xg for 10 minutes. Be sure that a counterbalance tube is used in the centrifugation.

4) Separate the pellet from the supernatant. Dab any excess supernatant with a cloth such as Kimwipe to prevent contamination of the two fractions. Reserve both fractions.

5) Resuspend the pellet in enough water to create 5 mL of solution.

6) Measure the volume of supernatant from Step 4. Calculate the amount of ammonium sulfate that must be added to that volume to create a 40% ammonium sulfate solution. (Refer to Table 6.1 and Section 6.5.)

7) Repeat Steps 1-4 with the amount of solid ammonium sulfate determined in Step 6.

8) Resuspend the pellet in enough water to create 5 mL of solution.

9) Measure the volume of supernatant from Step 7. Calculate the amount of ammonium sulfate that must be added to that volume to create a 60% ammonium sulfate solution.

10) Repeat Steps 1–4 with the amount of solid ammonium sulfate determined in Step 9.

11) Resuspend the pellet in enough water to create 5 mL of solution.

12) Measure the volume of supernatant from Step 10. Calculate the amount of ammonium sulfate that must be added to that volume to create an 80% ammonium sulfate solution.

13) Repeat Steps 1–4 with the amount of solid ammonium sulfate determined in Step 11.

14) Resuspend the pellet in enough water to create 5 mL of solution.

15) Measure the volume of supernatant from Step 13. Calculate the amount of ammonium sulfate that must be added to that volume to create a 100% ammonium sulfate solution.

16) Repeat Steps 1–4 with the amount of solid ammonium sulfate determined in Step 15.

17) Resuspend the pellet in enough water to create 5 mL of solution.

## Desalting of the Samples

1) Place each of the resuspended pellets into prepared dialysis tubing or centrifuge filtration devices. Make sure that the MWCO for the tubing or filter is appropriate for the proteins in your mixture.

2) Dialyse the samples overnight against the Tris buffer.

3) Alternatively, concentrate the sample with the centrifuge filtation device using manufacturer's specifications. After concentration, dilute the concentrate in enough Tris buffer to create a 5 mL solution. Repeat the centrifugation. Following centrifugation, add enough Tris buffer to create a 5 mL solution. Repeat the centrifugation. Repeat this washing and desalting at least 3 times. Dilute the sample to 5 mL with Tris buffer.

4) Store samples in refrigerator until next lab period.

## Characterization of the Fractions

1) Using procedure(s) in Experiment 5, determine the concentration of protein in each fraction. Determine the concentration of protein in the original sample.

2) Determine the composition of each fraction using SDS-PAGE. Include a lane with the original protein mixture on the gel. See Experiment 13 for experimental details. **Note –** for SDS-PAGE gels, the maximum concentration of protein that should be loaded per lane is 2 mg/mL of protein. If this is exceeded, individual bands will be difficult to observe or the sample may diffuse into neighboring lanes.

## Optional Additional Explorations

Design experiments to investigate the effects of salts other than ammonium sulfate, organic solvents, TCA, or PEG on the solubility of proteins. Compare the results of these experiments with those from the ammonium sulfate precipitation. Prior to any investigations, seek approval for the procedures from your instructor.

## Questions

1) What was the actual order of solubility in the experiment? Did it agree with your prediction? Explain.

2) What concentration range of ammonium sulfate is needed to precipitate each protein?

3) Determine the amino acid composition of each protein in the mixture used in the experiment. Does the amino acid composition have any effect on the solubility of a protein? Is there any other structural effect of the proteins that seems to have an impact on its solubility?

4) If you have conducted some of the other experiments in this text using the same proteins, compare your findings. Does one method provide information that differs from the others? Does one method separate the proteins better?

## References for the Experiment

Bollag, D. M., Rozycki, M. D., and S. J. Edelstein. *Protein Methods*. 2nd Edition. New York, NY: Wiley and Sons, Inc., 1996.

Spadaro, A. C. C., et al. Salt Fractionation of Plasma Proteins. *Biochem. Mol. Biol. Educ.*, 31, 2003: 249-252.

# Experiment 17. Extraction and Characterization of Bacterial DNA

## Purpose of the Experiment

In this experiment, DNA will be isolated from bacterial cells. Alternatively, DNA could be isolated from a plant source, such as an onion. After isolation, the concentration and purity of the DNA will be determined using several experimental methods.

## Prelaboratory Assignment

1) In the DNA isolation, what lyses the cell and the nucleus? What precipitates the DNA?

2) In the past, ethidium bromide was frequently used to stain DNA in gels. Why are alternative stains used in place of ethidium bromide? How does Fast Blast stain the DNA? Could the pH of the electrophoresis buffer have an impact on the binding of Fast Blast to the DNA?

## Materials

Selected liquid growth media for the bacteria – one of the most common is LB media (2.5 g tryptone, 12.5 g yeast extract, 1.25 g NaCl, 0.25 mL 1 M NaOH in 250 mL of solution; autoclave before using)

Agar plate containing the appropriate media for the bacteria Bacterial cells (for 10 LB plates, 2.5 g tryptone, 12.5 g yeast extract, 1.25 g NaCl, 0.25 mL 1 M NaOH, 3.75 g agar, in 250 mL of solution; autoclave, allow to cool to 50°C, pour into Petri dishes, remove air bubbles by quickly flaming the surface with a Bunsen burner, and allow to cool to hardness. Store in the refrigerator.)

Saline-EDTA solution (0.15 M NaCl and 0.1 M EDTA, pH 8)

25% SDS in water

Lysozyme solution (10 mg/mL in water)

5 M sodium perchlorate

Chloroform:isoamyl alcohol (24:1)

95% ethanol

Tris buffer (0.01 M Tris-HCl and 0.05 M NaCl, pH 7.5)

Quartz cuvettes

Spectrophotometer

Eppendorf tubes

Materials for DNA concentration determination (see Experiment 6)

Test tubes

DNA standard solutions (100 ng/mL and 2 mg/mL)

Electrophoresis apparatus

Agarose

1X TBE buffer (89 mM Tris base, 89 mM boric acid, and 2 mM EDTA, pH 8.0)

DNA electrophoresis loading buffer (30% glycerol, 0.25% (w/v) bromophenol blue, 0.25% (w/v) xylene cyanol)

DNA Ladder

Ethidium bromide or alternative DNA stain

## Safety

Working with bacterial cells presents a potential biohazard. Do not pipet solutions by mouth. Wear gloves when handling the bacterial cells, and wash your hands thoroughly with soap and water. Ethidium bromide is a suspected carcinogen. Wear gloves when handling ethidium bromide solutions. Wear goggles to protect your eyes when using the UV transilluminator. Dispose of the solutions at the end of the experiment in the proper manner, as indicated by the instructor.

## Experimental Methods

*Culturing Bacterial Cells* – required in advance of isolation of DNA. Your instructor or laboratory aide may complete this in advance of the laboratory period.

1) Streak an agar plate with a sample of bacteria on a sterile wire loop.

2) Allow the plate to incubate at 37°C overnight.

3) Using a sterile wire loop, select a colony from the plate and inoculate 100 mL of liquid growth media with the colony. Repeat with a second colony.

4) Allow samples to grow at 37°C overnight.

5) Centrifuge the sample at 10,000 g for 15 minutes.

6) Remove the supernatant. The pellet is the wet-packed bacterial cells. The sample can be frozen at this point or used to isolate the DNA immediately.

*Isolation of DNA*

1) Suspend 2–3 g of wet-packed bacterial cells in 25 mL of saline-EDTA solution in a 125 mL Erlenmeyer flask.

2) Add 1.0 mL of lysozyme solution, and incubate the mixture at 37ºC for 30–45 minutes.

3) To complete the cellular lysis, add 2.0 mL of 25% SDS, and heat the mixture to 60ºC for 10 minutes.

4) To avoid excessive foaming, stir the solution gently. The nucleic acid should be released from the cells, and the solution should increase in viscosity and cloudiness.

5) Allow the mixture to cool to room temperature, or cool the flask under cold running water.

6) Add 9.0 mL of sodium perchlorate. Mix well, but gently.

7) Add a volume of chloroform:isoamyl alcohol (24:1) that is equal to the total volume of the extraction mixture.

8) Shake the mixture in a separatory funnel or stoppered flask for 10–15 minutes.

9) Centrifuge the emulsion for 5 minutes at 10,000 xg. Three layers should be present: a bottom organic layer, a middle layer of denatured proteins at the interface, and an upper aqueous layer containing the nucleic acids.

10) Carefully transfer the aqueous layer to a 125 mL beaker.

11) Precipitate the nucleic acids by carefully layering about 2 volumes of 95% ethanol over the aqueous phase. Pour the ethanol down the side of the beaker. Mix the two layers very gently using a circular motion with a stirring rod, and "spool" all the fibers of nucleic acid onto the rod.

12) Press the spooled nucleic acid against the inside of the beaker to remove solvent.

13) Dissolve the nucleic acid in 10–15 mL of Tris buffer. If necessary, save the DNA in the refrigerator for other studies.

*Characterization of DNA*

1) Using procedure(s) from Experiment 6, determine the concentration of the DNA in the sample.

2) Pour a 1% agarose gel in 1X TBE buffer. Based on manufacturer's information, determine the volume of agarose solution required to create the gel. Measure the appropriate amount of agarose and place in an Erlenmeyer flask with the determined volume of TBE buffer. Gently heat the solution over a hotplate, in a microwave, or in an autoclave until the agarose melts and a homogenous solution is created. Avoid boiling or charing the agarose.

3) Cool the solution to 55°C and pour into the gel cast. Remove any bubbles, and insert the comb.

4) Allow the gel to harden. Once hardened, remove the gel comb and place the gel into the electrophoresis chamber. Cover the gel with TBE buffer.

5) Mix the DNA sample with 0.2 volumes of loading buffer. If necessary, mix the DNA ladder with 0.2 volumes of loading buffer (some manufacturers prepare DNA ladder samples containing loading buffer).

6) Slowly load the samples into the wells in the agarose gel using a micropipetter. Load the DNA standards (ladder) on both the right and left sides of the gel.

7) Close the electrophoresis apparatus. Apply a voltage of 1–10 V/cm until the bromophenol blue has migrated within 1 cm of the end of the gel.

8) Turn off the power, disconnect the leads, and remove the gel from the electrophoresis chamber.

9) Stain the gel by soaking in TBE buffer containing 0.5 µg/mL ethidium bromide (or other appropriate DNA stain) for 10–30 minutes.

10) Visualize the DNA bands by illuminating the gel with UV light using a transiluminator.

11) Document the gel using photography.

## Questions

1) What is the concentration of isolated DNA? What is the approximate size of the DNA isolated from the bacteria?

2) Is the isolated sample of DNA pure? What evidence do you have to support your answer?

## References for the Experiment

Boyer, R. *Modern Experimental Biochemistry*. 3rd Edition. San Francisco, CA: Benjamin Cummings, 2000.

Sanbrook, J., and D. W. Russell. *Molecular Cloning. A Laboratory Manual.* 3rd Edition. Cold Spring Harbor, NY: Cold Spring Harbor Laboratory Press, 2001.

# Separation of Biological Molecules by Chromatography

Chromatography is a powerful analytical separation technique. It was invented and named by Mikhail Tswett. He was a botanist who separated plant pigments that appeared as colored bands. This accounts for the name he chose: *chroma* (Greek) meaning "color" and *graphein* meaning "to write." While this reference to color only applies to the first separations of colored substances by Tsweet using chromatography, today chromatography is used to separate colored and uncolored species alike and is utilized in most scientific disciplines. Box 7.1 highlights some of the uses of chromatography in forensics.

---

Science in Action

### Box 7.1 - Use of Chromatography in Forensic Applications

Chromatography is a frequently used tool by forensic scientists. Illegal drugs and other controlled substances are routinely screened using thin layer chromatography (TLC) and are identified using techniques such as gas chromatography with flame ionization detection (GC/FID) or mass spectrometry (GC/MS), capillary electrophoresis (CE), or high-performance liquid chromatography (HPLC). In addition, agents working for the Drug Enforcement Agency (DEA) analyze the adulterants, diluents, and drug precursors to determine the source of drugs seized from clandestine laboratories. An example of studies conducted by the DEA involved analysis of 54 different types of Ecstasy tablets seized in Japan in 2002. GC/MS and HPLC were used to determine the active ingredients in each tablet. It was found that the amount of the active ingredient 3,4-methylene-dioxymethamphetamine varied from 15–64% of the tablet, and the amount of adulterants or diluents in the tablets varied by comparable amounts. Using similar methods, explosives and fibers (mainly the dyes and contaminants in the fibers) are analyzed by chromatographic techniques as evidence in criminal cases.

---

## 7.1 GENERAL PRINCIPLES OF CHROMATOGRAPHY

The different forms of chromatography vary considerably, but, in general, chromatography occurs because of the distribution of a substance between a stationary phase and a mobile phase. One way to classify chromatography depends on the form of the stationary phase. Column chromatography involves passing the mobile phase (under pressure or the influence of gravity) through a narrow tube that is either packed or coated with the stationary phase. In planar chromatography, the stationary phase is either paper or a thin film on a solid support. The mobile phase travels through the stationary phase by capillary action or gravity influence. A more common method in which chromatography is classified is based on the physical state of the stationary and mobile phases used. Table 7.1 shows different types of chromatography depending on the mobile and stationary phases. A mobile phase can be either gas, liquid, or supercritical fluid that moves past the stationary phase, an immiscible fixed phase. This chapter will only focus on gases and liquids and will not discuss supercritical fluids. For the former, the stationary phase is typically a solid or a liquid.

Table 7.1 – Classification of Chromatographic Methods.

| Type | Mobile Phase | Stationary Phase | Type of Equilibrium |
|---|---|---|---|
| Liquid Chromatography (LC) | | | |
| Liquid-liquid | Liquid | Liquid adsorbed on a surface | Partition between liquids |
| Liquid-bonded phase (affinity chromatography) | | Molecules bonded on a surface | Partition between liquid and bonded species |
| Liquid-solid | | Solid | Adsorption |
| Ion exchange | | Ion exchange resin | Ion exchange |
| Size exclusion or gel-filtration | | Porous solid support | Partitioning |
| Gas Chromatography (GC) | | | |
| Gas-liquid | Gas | Liquid adsorbed on a surface | Partition between liquids |
| Gas-bonded phase | | Molecules bonded on a surface | Partition between liquid and bonded species |
| Gas-solid | | Solid | Adsorption |

### *Plate Theory*

In chromatography, the compounds separate by establishing an equilibrium between the stationary and mobile phases. This theory of partitioning between the stationary and mobile phases was developed by Archer John Porter Martin and Richard Laurence Millington Syrge. Consider an equilibrium in which the analyte is partitioning between the stationary and mobile phases:

$$A_{\text{mobile phase}} \rightleftharpoons A_{\text{stationary phase}}$$

The equilibrium between the two phases can be described as

$$K = \frac{\text{concentration in stationary phase}}{\text{concentration in mobile phase}}$$

The difference in the equilibrium constants or partitioning coefficients among analytes results in the resolution of a mixture. A compound with higher affinity for the stationary phase will have a higher K and will be retained longer on the column than a compound with a higher affinity for the mobile phase. As a result, compounds move in discrete bands or zones off of the stationary phase. K can be assumed to be independent of initial concentration and can change if experimental conditions, such as temperature, are changed.

Martin and Syrge defined a plate as the equilibrium existing between the stationary and mobile phases. No real plates exist in a column but are a useful concept to understand how separations occur. The greater the number of plates in a column, the better the separation of the analytes. The height of a plate (H) can be determined from the length of the column (L) and the number of theoretical plates (N):

$$H = L/N$$

Assuming that the peaks are gaussian, the number of theoretical plates can be determined from the retention time ($t_r$) and the width at the base of the peak (w):

$$N = 16\left(\frac{t_r}{w}\right)^2$$

An example of column chromatography is depicted in Figure 7.1. A sample containing two analytes, A and B, is loaded onto the top of the stationary phase. The mobile phase flows through the column. As the mobile phase passes over the stationary phase, the analytes are separated based on the mobility of each molecule, which is dependent on the flow rate of the mobile phase and the partitioning coefficient (K). Additional mobile phase continues to carry the molecules down the column, and more partitioning occurs between the stationary and mobile phases. As shown in Figure 7.1, analyte A has a lower K, signifying a lower affinity for the stationary phase, and moves at a faster rate through the column than analyte B.

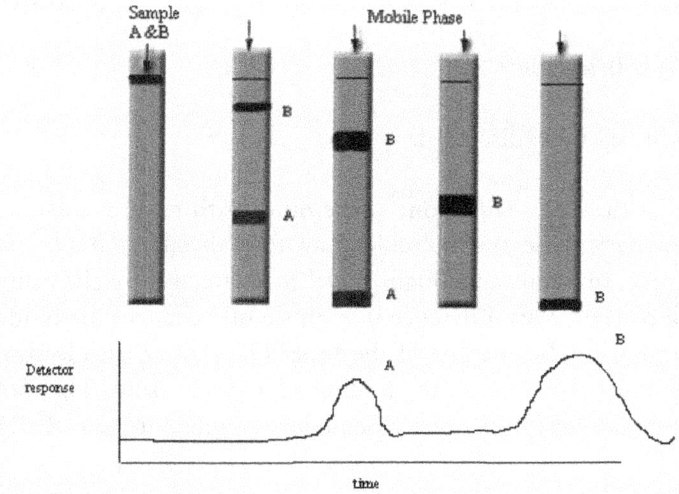


Figure 7.1 – Schematic representation of separations in column chromatography. Compound A has a lower affinity for the stationary phase and elutes first, while Compound B is retained longer on the column.

### Chromatograms

Frequently, chromatography columns are connected to a detector. As the analyte elutes from the column, its presence is noted by the detector, and a signal is observed (i.e., a series of peaks is generated, as shown in Figure 7.1). A plot of detector signal versus time is referred to as a chromatogram. The time between the injection (time zero) and the detection of the compound (appearance of a peak) is called the retention time ($t_r$) and is based on the affinity for the stationary and mobile phases. In some forms of chromatography, such as TLC, scientists are also often interested in identifying the time required for the mobile phase to travel through the column ($t_m$). The retention time or retention factor can often be used to identify a compound. The area under the peak is related to the concentration of the substance. Assuming that the peak is perfectly triangular (although this rarely occurs), the area of the peak is calculated by multiplying the height of the peak by the width at half the height. The areas of two or more peaks can be used to determine the percent composition of a mixture. Because of the information provided by the retention time and by the peak area, chromatography can be easily used to obtain both qualitative and quantitive information about a mixture simultaneously.

### *Peak Broadening*

The plate model assumes that a very rapid equilibrium is established between the stationary and mobile phases. In the presence of a constantly moving mobile phase, a true equilibrium is not reached between the two phases. In reality, two factors affect the efficiency of the separations using chromatography. First, the analytes must have distinctly different partition coefficients so that good separation is achieved. This factor can be controlled through the proper selection of mobile and stationary phases to ensure different relative affinities.

Second, dispersion or band broadening can impact separation. van Deemter developed an equation and theory that accounts for the effect of band broadening on the separation. The van Deemter equation shows the dependence of plate height (H) on several factors associated with zone broadening (A, B, and C) as well as on the average velocity of the mobile phase ($u$):

$$H = A + B/u + Cu$$

Factors A, B, and C will be briefly discussed.

Factor A is referred to as the Eddy Diffusion, or the multi-path term. Depending on the nature and packing of the stationary phase, the molecules may travel through the stationary phase by different paths. For example, one molecule might travel in a directly straight route, while another may travel in a zig-zag pattern. This difference in path slightly changes the residence time of the molecule and would result in a broadening of the zone. The shorter path leads to a shorter residence time, and these molecules elute on the leading edge of the band. The molecules traveling through a longer path remain in the stationary phase longer and elute near the tailing end of the band.

Factor B is the longitudinal diffusion term. The solute molecules may diffuse slightly from the center of the zone to a neighboring part of the stationary phase. Molecules traveling ahead of the center of the band elute faster than the average molecules. Other molecules will remain behind and elute later than the majority of the particles. This difference in diffusion through the stationary phase also contributes to the band broadening. This factor is inversely proportional to the mobile phase velocity. If the flow rate is higher, the analyte spends less time in the column and less diffusion can occur.

Variable C is related to the mass transfer between the mobile and stationary phases. This factor is directly related to the flow rate. Analytes will equilibrate between the two phases, and this equilibration requires time to be established. At a given flow rate, some of the analyte will be able to transfer from the mobile phase into the stationary phase, but not all of the molecules with the same affinity will be able to transfer into the stationary phase. Those molecules that did not enter the column material will be pushed ahead, and, as a result, band broadening occurs. Likewise, some material will be retained, not entering the moving mobile phase, and tailing occurs.

Peak broadening can lead to an overlap of bands and result in poor resolution. One way to control peak broadening is to optimize the chromatography condition, such as column size, diameter of the particles in the stationary phase, sample size, the flow rate of the mobile phase, and the temperature.

This introduction to chromatography is meant to provide a basic understanding of the theory behind this technique. Some of the forms of chromatography used in biochemical application will be discussed in the remainder of this chapter. If desired, more details on chromatography can be found in the references. (See References and Further Readings.)

## 7.2 GAS CHROMATOGRAPHY (GC)

Gas chromatography is an analytical technique in which gaseous samples are separated while carried by a gaseous mobile phase. Gas chromatography can provide both qualitative and quantitative information for the components of the sample. The samples must be gases or volatile liquids that can be vaporized into gases. The gaseous compounds are separated through partitioning between the gaseous mobile gas and the stationary phase, either a liquid or a solid. The most common form of GC is gas-liquid partition chromatography in which the stationary phase is a nonvolatile liquid polymer coated on an inert solid support. Gas-solid adsorption chromatography utilizes a porous solid stationary phase on which the sample molecules can be adsorbed.

The stationary phase is present in a tubular column and attracts organic molecules to varying extents depending on the intermolecular forces. The mobile phase is a carrier gas, typically helium, nitrogen, argon, or hydrogen gas. The compounds applied to the column equilibrate according to their distribution coefficient (K) between the stationary phase and the mobile phase. After equilibration, the molecules in the mobile phase continue to move down the column by re-equilibrating according to their distribution coefficients. This continues until separation is achieved, and the compound with a lower affinity for the stationary phase is eluted first.

Gas chromatographs are fairly simple instruments. The sample to be analyzed is injected into the mobile phase through a heated injector port, where the sample is vaporized. The sample is then swept into the column containing the stationary phase by the carrier gas. The mobile phase moves through the column, and the mixture is separated as described above. Finally, the carrier gas flows through a detector, which senses the presence of the compounds and sends a signal to a recorder. The result is a chromatogram.

The detector and injector are maintained at a higher temperature (25–50°C above the boiling point of least volatile component of sample) than the column so that all of the components are in the gas phase as they pass through these chambers. The column temperature must be high enough for the components to have sufficient vapor pressure to elute in a reasonable time, but it does not need to be above the boiling point of all the components. The rule of thumb is that a temperature slightly above the average boiling point of the sample results in elution times of 2–30 minutes. Lower temperatures provide good resolution but result in increased elution times.

Column ovens can operate at one set temperature (isothermally) or can be programmed to change temperatures at predetermined rates. Isothermal separations work well for mixtures that contain substances with boiling points that differ by no more than about 50°C. Complex mixtures are better separated using thermal gradients.

There are several types of columns depending on the type of instrument and detector being used. Packed columns contain a finely divided, inert, solid support material that is coated with a liquid stationary phase. Capillary columns are thin-walled, narrow, tubular columns in which either the walls are coated with the liquid stationary phase or the walls are lined with a thin layer of support on which the liquid adheres. Capillary columns are much more efficient than packed columns. The structure of the stationary phase affects the retention times of the compounds in a mixture. Typical stationary phases are polydimethyl siloxane, poly(phenylmethyl) siloxane, polyethylene glycol, or polyester polymers.

The detector senses when a compound is exiting the column and sends a signal to the recorder or, with more sophisticated instruments, to a computer. A thermal conductivity detector (TCD) is one of the earliest detectors developed for GC and is the most common type on inexpensive instruments. It is rugged, has universal selectivity, and can detect approximately 10 nanograms of sample per mL of gas. The TCD requires that the carrier gas stream be split into two parts, one for reference and one for the sample that passes through the column. The TCD contains two hot filaments (composed of platinum, gold, or tungsten). As the temperature increases, the electrical resistance of each filament increases. Helium, which has a high thermal conductivity and is good at dissipating the heat created by the filaments, is the carrier gas of choice for a TCD. As long as helium is flowing at a constant rate over both filaments, the resistance is constant in both sides of the detector, and a constant signal is sent to the recorder. When a compound emerges from the column, the thermal conductivity of that gas stream decreases, so the rate at which that filament is cooled by the gas stream decreases. That filament becomes hotter, its resistance increases, the current decreases in that arm of the detector, and a change in signal is sent to the recorder.

A flame ionization detector (FID) is the most widely used detector. It detects most organic compounds and is much more sensitive than a thermal conductivity detector. It can detect as little as 0.1 picogram of sample per mL of gas. The effluent, the combination of the mobile phase and analyte exiting from the column, is mixed with $H_2$ and air, and is burned in a flame inside the detector. Most organic compounds undergo pyrolysis and produce electrons and ions. The collector electrode collects the ions that are produced, and the change in current due to the presence of the ions is measured by the electrode and transmitted to a recorder.

There are other types of detectors, including the mass selective detector (MSD, mass spectrometer) and the electron capture detector (ECD). See Chapter 3 for details on mass spectrometry. The electron capture detector is frequently used for environmental samples (such as pesticides or polychlorinated compounds) and detects the presence of halogen-containing compounds. The effluent from the column is passed over a radioactive beta-emitter, which causes the ionization of the carrier gas. The ionization process produces a spray of electrons, which are detected by a pair of electrodes. As an organic compound is eluted from the column, it reacts with or captures the electrons, the current decreases, and a signal is recorded by the detector.

## 7.3 THIN LAYER CHROMATOGRAPHY (TLC)

Thin layer chromatography is a simple analytical technique for separating and identifying the compounds in a mixture. It is often used in by forensic scientists as a rapid screening tool for drug detection. (See Box 7.1.) TLC is one type of liquid-phase chromatography in which moderately volatile or nonvolatile substances are separated based on differential adsorption on a solid stationary phase. The mobile phase in TLC is a liquid, either a pure solvent or a mixture of solvents or buffer. TLC is a very sensitive technique and requires only micrograms of material ($10^{-9}$ g can be detected). The introductory description of chromatography mainly described column chromatography. Although many of the principles are similar, the stationary phase in TLC is not in a column but is present as a thin layer on a solid support, such as glass, plastic, or aluminum.

Samples are spotted on a TLC plate, a sheet of plastic, or glass coated with a thin layer of adsorbent. The plate is then placed in a covered jar perpendicular to a shallow layer of solvent. The solvent rises up the plate by capillary action, and the sample partitions between the mobile phase and the stationary phase. Components that strongly adhere to the stationary phase spend less time in the mobile phase and migrate up the TLC plate more slowly.

The most common adsorbents are alumina and silica gel. Chromatographic paper (cellulose) is also frequently used in biological applications. Alumina and silica are polar inorganic polymers with oxygen atoms and hydroxyl groups on the surface, whereas cellulose is a polar polysaccharide containing polyhydroxyl groups. These polar stationary phases strongly attract polar substances. Nonpolar compounds are only weakly attracted to these polar adsorbents. In general, nonpolar compounds move faster than polar compounds on TLC. The degree of separation and movement up the plate will depend on the nature of the solvent used for development.

When handling the TLC stationary phase, do not touch the surface of the plate or paper. Fingerprints can interfere with development, obscure spots, or be mistaken for spots. Wear gloves, and handle the plate by the edges to prevent damage to the stationary phase. Using a pencil, make a small mark on one side of the stationary phase at least one centimeter from the bottom (the starting line). Small samples are pipeted onto the plates or spotted by touching fine capillaries dipped into the samples onto the plates along the starting line about ¼ of the way across the plate. Allow the sample to form a small spot about 1 mm in diameter. The diameter of the spot should be kept as small as possible to minimize diffusion effects. After the solvent has evaporated, you may add more sample to the same spot to increase the concentration as necessary.

TLC plates are developed in a chromatography jar with a tight lid or covered with a glass plate. Place about 0.5 cm of the developing solvent or buffer in the bottom of the jar. Cut a piece of filter paper to line about one-half of the inside wall of the jar. Allow the developing chamber to stand for 5–10 minutes to saturate the air in the jar with solvent vapors. Keep the jar closed as much as possible.

Put the plate into the chamber with the stationary phase face up (if appropriate); make sure that the spots do not dip into the solvent. The solvent will flow up the plate by capillary action, and the sample will equilibrate between the stationary phase and the mobile phase. When the solvent level is about 1 cm from the top of the plate, remove the plate from the chamber, and mark the solvent front with a pencil. Do not allow the solvent front to reach the top of the plate.

Most organic compounds are colorless and a visualization technique is necessary to see the spots on the plates. A variety of methods have been employed to visualize the spots. Most silica gel plates are impregnated with a fluorescent dye, and, under 254 nm UV light, the dye fluoresces with a greenish-yellow glow. Many organic compounds quench the fluorescence and appear as black spots. Fully aliphatic compounds do not show up well by this technique. Some organic compounds, mostly conjugated aromatics, fluoresce on their own and appear as bright spots under UV light.

Iodine vapor can also be used to visualize spots. Most organic compounds adsorb iodine vapor, and a brown or purple spot will form where the organic compound is present on the plate. The plate is placed face down in a jar containing iodine crystals and allowed to stand for 15–20 minutes. Fully saturated compounds with no oxygen or nitrogen do not show up well with this technique. After removing the plate from the iodine, it is necessary to circle the spots because the iodine will sublime, leaving the spot colorless again.

Proteins and amino acids can be visualized by treating the stationary phase with compounds such as ninhydrin. Ninhydrin reacts with the amino group on amino acids and proteins and forms the purple compound shown in Figure 7.2. The only exception to this reaction is proline, which reacts with ninhydrin to form a yellow compound. Care should be taken when using ninhydrin to avoid contact with skin, as the proteins in skin will react, coloring your skin purple until it is sloughed off. Because of this high affinity for proteins, peptides, and amino acids in the skin, ninhydrin is frequently used by forensic scientists to detect fingerprints on surfaces.

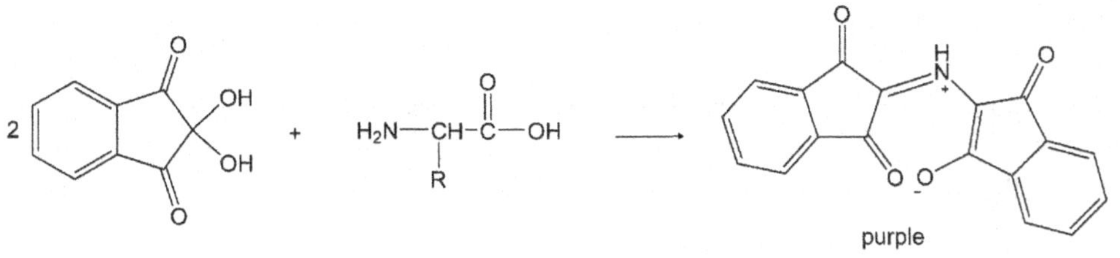

Figure 7.2 – The reaction of ninhydrin with an amino acid.

After a plate is developed and the spots have been visualized, the retention values ($R_f$) for each substance should be determined. On TLC, an $R_f$ value is equal to the $d_{sample}/d_{solvent}$ for each spot, where $d_{sample}$ is the distance traveled by a spot, by measuring from the origin to the center of the spot on the plate or to the leading edge of the spot, and $d_{solvent}$ is the distance the solvent traveled from the origin. Assuming that the stationary phase and conditions are uniform (i.e., same solvent, same temperature, same thickness stationary phase), the $R_f$ value for a compound should be the same from one plate to another.

## 7.4 HIGH-PERFORMANCE LIQUID CHROMATOGRAPHY (HPLC)

High-performance liquid chromatography is a type of liquid chromatography in which the mobile phase is passed through the column using pressurized pumps rather than classic gravity-flow. Originally, the technique was called high pressure liquid chromatography because of the pressurized flow of the mobile phase. Pressure, however, is not the primary factor influencing the high degree of separation achieved in HPLC. Pressure is necessary to pass the mobile phase through a column at the specified flow rate, but differences between HPLC and classic column chromatography also include the use of smaller-diameter columns, column packing with smaller particles, and smaller sample sizes, which of all contribute to HPLC's rapid analysis and high resolution. For this reason, many scientists today refer to the technique as high-performance liquid chromatography.

The sample, which is dissolved in solution, is injected into an injection port. The sample is passed through the column, where separation is achieved by the equilibrium. The column is connected directly to a detector, and the effluent is passed through a detector. Separations can be achieved isocratically or through gradient elution. In isocratic separations, the composition of the mobile phase does not change. In a gradient elution, a pre-programmed change in the mobile phase composition occurs. The change in the solvent composition can occur either step-wise or in a linear fashion. The choice of elution depends on the type of separation to be achieved and the chemical makeup of the analytes.

The stationary phase provides the basis for the separation in LC. The stationary phase in HPLC consists of material that separates the compounds based on intermolecular forces. The phase may be charged, hydrophobic, porous, or contain an affinity ligand. When a sample is applied to the column, it equilibrates or partitions between the stationary phase and the mobile phase. Substances with a higher affinity for the stationary phase will elute later than weakly bound substances. The most commonly used types of HPLC for biochemical purposes (involving different stationary phases) will be described in Sections 7.6–7.9. These techniques can either be used in HPLC systems or gravity-flow applications.

In addition to the chemical nature of the stationary phase, the polarity of the stationary and mobile phases has an effect on the separations. Normal phase chromatography utilizes a stationary phase that is polar, for example silica gel, and a mobile phase that is nonpolar, such as hexane. Based on intermolecular forces, nonpolar substances will have a higher affinity for the mobile phase and be retained less than the polar substances attracted to the stationary phase. Reversed-phase chromatography is the opposite of normal phase. The stationary phase is nonpolar, such as a C-8 column that contains an octyl chain or a C-18 column with an octadecyl hydrocarbon chain, and the mobile phase is polar, such as water, methanol, or acetonitrile. In this form of chromatography, the nonpolar materials are retained longer on the column, while the polar molecules elute first. The degree of separation in both forms are, therefore, influenced by the polarity of both phases; changing the stationary or mobile phase will change the resulting separation.

The detector and data analysis systems are one of the key components in a HPLC system. As described in Section 7.2, a detector senses the presence of an eluting sample and sends a signal to the integrator or computer to produce a peak on the chromatogram. There are many types of detectors employed with HPLC, and some of the more common ones include ultraviolet-visible (UV-Vis), fluorescence, and electrochemical detectors. It is becoming more common to couple HPLC separations with more advanced spectroscopic techniques such as MS or NMR. For more information on spectroscopy, please refer to Chapter 3.

Ultraviolet-visible detectors are the most-utilized HPLC detectors. As described in Chapter 3, most organic compounds absorb light in the UV-Vis range and can be detected using this technique. The sample can be detected using a fixed wavelength UV-Vis detector, in which only one wavelength is measured, usually 254 nm, or a variable wavelength detector, in which only one wavelength is measured at a time, but can be changed. A third type of UV-Vis detector uses a diode array to measure several wavelengths simultaneously.

Fluorescence detectors are 10–1000 times more sensitive than UV-Vis detectors. These detectors sense the fluorescence of the eluted molecules. The detectors are generally set to one excitation and one emission wavelength but can be varied by the controller. These detectors also tend to be more selective, as they are measuring specific fluorescent species in the sample.

Electrochemical detectors measure the current resulting from oxidation-reduction reactions occurring between the eluting sample and an electrode in the detector. These detectors are highly sensitive and selective. The level of current produced is proportional to the amount of substance eluted from the column so, as with UV-Vis and fluorescence detectors, electrochemical detectors can also be used qualitatively and/or quantitatively. Care needs to be taken with electrochemical detectors to ensure that oxygen, metal, or halogen contaminants are avoided in the mobile phase or sample, as these will easily be detected electrochemically.

## 7.5 COLUMN CHROMATOGRAPHY

Liquid chromatography is the general term for a wide range of analytical techniques that permit the separation of compounds based on different partitioning between a stationary phase and a liquid mobile phase. This chapter has already explored two applications of liquid chromatography, TLC and HPLC. In addition to these forms of liquid chromatography, separations can also be carried out in glass or plastic columns packed with the chromatographic support. The liquid phase passes through the column by gravity rather than by a pump. Often, peristaltic pumps will be used to deliver the mobile phase to a reservoir on top of the column, but no pressure is applied to increase the flow through the column.

Several problems can occur with packed columns and should be avoided. When the solid support is poured into the column, air bubbles or cracks in the matrix must be avoided. If not removed, these holes can create pockets in which the molecules to be separated can reside, and inefficient or insufficient separations can occur. If air bubbles or cracks are observed, the column needs to be repacked. Air bubbles frequently occur if the solid support was not degassed before pouring the gel or if the column is subjected to a temperature change. If the column is allowed to run dry, cracks or pockets can occur. Insoluble particles in samples applied on the column can often cause poor resolution of molecules eluted from the column and can slow or stop the flow of the mobile phase through the column.

Most columns can be stored between uses. For long-term storage, the column material should be stored in 0.02% sodium azide at 4°C or as suggested by the manufacturer to prevent microbial growth. The presence of microbes in the column matrix can limit the column flow, cause a poor recovery of material from the column, or decrease resolution.

## 7.6 ION EXCHANGE CHROMATOGRAPHY

In ion exchange chromatography, the resin in the stationary phase contains charged groups on the surface. Some common resins used in biochemical experiments are shown in Table 7.2. The choice of the resin depends on the size and ionic properties of the molecules to be separated. For protein separations, if a protein is more stable at a pH lower than the protein's pI, the protein will be positively charged at that pH and should be separated using cationic exchange chromatography. If a protein is more stable at a pH higher than the protein's pI, anionic exchange chromatography is employed. The molecular weight of the protein can also influence the choice of a resin. Proteins that are somewhat small in size (MW 10,000–1,000,000 g/mol) use DEAE columns. Larger proteins are better separated on Sephadex columns.

In anion exchange chromatography, for example DEAE-Sephadex, the stationary phase is positively charged (i.e., contains an quarternary ammonium salt). The quarternary ammonium ion is bound to the resin core and is associated with an anion. The anion (frequently a Cl$^-$) can be exchanged for other anions. This principle is very useful for separating polypeptides, nucleic acids, and amino acids. These molecules contain charged functional groups of differing electrical charges. On an anion exchange column, positively charged molecules will be eluted first, as they are repelled away from the stationary phase. The neutral molecules will be separated based on hydrophobicity. The more hydrophobic substances will interact with the core of the resin used in the stationary phase and will be retained slightly. Finally, negatively charged substances will be retained through ionic interactions with the stationary phase.

Table 7.2 – Typical Ion Exchange Resins Used in Biochemical Applications.

| **Anion Exchange** | | | |
|---|---|---|---|
| **Type** | **Functional Group** | **Resin Core** | **Exchange Type** |
| AG 1 | Tertiary amine | Styrene divinylbenzene | Strong |
| AG 2 | Tertiary amine | Styrene divinylbenzene | Strong |
| AG 4-X4A | Tertiary amine | Acrylic | Weak |
| Aminoethyl-Cellulose | Aminoethyl | Cellulose | Weak |
| DEAE-Cellulose | Diethylaminoethyl | Cellulose | Weak |
| DEAE-Sephadex | Diethylaminoethyl | Dextran | Weak |
| DEAE-Sepharose | Diethylaminoethyl | Sepharose | Weak |
| QAE-Cellulose | Diethyl-(2-hydroxypropyl)aminoethyl | Cellulose | Strong |
| QAE-Sephadex | Diethyl-(2-hydroxypropyl)aminoethyl | Dextran | Strong |
| Q-Sepharose | Tetramethyl amine | Agarose | Strong |
| **Cation Exchange** | | | |
| **Type** | **Functional Group** | **Resin Core** | |
| Bio-Rex 70 (Dowex) | Sulfonic acid | Polystyrene | Weak |
| CM-cellulose | Carboxymethyl | Cellulose | Weak |
| CM-Sephadex | Carboxymethyl | Sephadex | Weak |
| CM-Sepharose | Carboxymethyl | Sepharose | Weak |
| S Sepharose | Sulphonate | Sepharose | Strong |
| SP-Sephadex | Sulphopropyl | Sephadex | Strong |

Cation exchange is very similar, except that the stationary phase is negatively charged, and cations can be changed. Initially, the column is often acid-washed and has $H^+$ associated with the negatively charged anionic groups chemically bound to the resin. The column is then washed with NaOH, which replaces the $H^+$ with $Na^+$ ions. The cation exchange column is used in either the $Na^+$ or $H^+$ form, depending on the application. $Na^+$ or $H^+$ will exchange for other cations, and, depending on relative affinity (based on charge on the ion), a mixture of cations will separate.

An ion exchange column should be prepared according to manufacturer's instructions. In ion exchange chromatography, a sample of molecules is dissolved in an appropriate buffer and is loaded onto the stationary phase. The sample is allowed to enter the column. Buffer is passed through the column to elute substances that are not adhered to the column. The adhered compounds are then eluted by passing a gradient of buffer with increasing salt concentration or change in pH through the column. Fractions are collected from the column and detected either using spectroscopy or other means.

## 7.7 GEL FILTRATION OR SIZE EXCLUSION CHROMATOGRAPHY

Gel filtration, otherwise called gel permeation, chromatography is an analytical technique that separates mixtures based on differences in size. The stationary phase in this method is a silica or polymer material that contains numerous pores with which the mixture components can interact. The pore size in the stationary phase is well-defined, and this size determines the molecular weights of the substances that can be separated. Molecules of small molecular weight and appropriate shape enter the pore and are trapped or retained on the column. Larger molecules are unable to enter the pore and are eluted from the column. Substances of intermediate size partially enter the pores and are retained for an intermediate amount of time. In this way, molecules of differing molecular weights can be separated. Included molecules (those interacting with the pores) are eluted by altering the buffer composition of the mobile phase or by simply passing mobile phase through the column. The small buffer molecules will interact competitively for the pores, and the included molecules will slowly elute from the column. In this manner, gel filtration separates proteins in order of large to small molecules.

Gel filtration matrixes are commercially available from Pharmacia, Bio-Rad, and Sepracor. One of the most common types of matrix used is Sephadex. Table 7.3 gives the properties of various types of Sephadex, including the fractionation range and size of the pores. If a protein of known molecular weight is to be purified or characterized, select a Sephadex gel on which the protein will elute midway in the fractionation range. If the molecular weight is unknown, initially select a broad fractionation range. If the protein elutes in the void volume (i.e., not retained in the column), repeat the experiment with a Sephadex with a higher exclusion limit. If the protein elutes after at least one bed volume (the volume of the stationary phase in the column) of buffer is applied, repeat with a gel matrix with a lower exclusion limit. Gel filtration is also used to separate nucleic acids from smaller molecules, such as nucleotides or other contaminants. Sephadex G-50 is frequently utilized for purifying DNA; the cut off for the pores is DNA that is larger than 80 nucleotides in length.

Table 7.3 - Properties of Sephadex.

| Type | Grade | Dry Bead Diameter (Ìm) | Fractionation Range (Molecular Weight) | | Bed Volume mL/g Dry Sephadex |
|------|-------|------------------------|------------------------|------------------------|------------------------------|
| | | | Globular Proteins | Dextrans | |
| G-10 | | 40-120 | <700 | <700 | 2-3 |
| G-25 | Coarse | 100-300 | 1,000-5,000 | 100-5,000 | 4-6 |
| | Medium | 50-150 | | | |
| | Fine | 20-80 | | | |
| | Superfine | 10-40 | | | |
| G-75 | Medium | 40-120 | 3,000-80,000 | 1,000-50,000 | 12-15 |
| | Superfine | 20-50 | 3,000-70,000 | | |
| G-100 | | 40-120 | 4,000-150,000 | 1,000-100,000 | 15-20 |
| | Superfine | 10-40 | 4,000-100,000 | | |
| G-150 | | 40-120 | 5,000-300,000 | 1,000-150,000 | 20-30 |
| | Superfine | 10-40 | 5,000-150,000 | | 18-22 |

The basic method for gel filtration involves preparing a column by creating a slurry of the dried gel in an appropriate buffer. The solution is degassed and poured carefully into a column to the desired bed volume. The column is then washed with two to three bed volumes of buffer. The sample is applied to the column, and, after the sample enters the medium, buffer is used to elute the molecules from the column. The sample applied should be less than 5% of the total bed volume. The molecules are eluted isocratically. Fractions equal to approximately 1–10% of the bed volume should be collected from the column. Collected fractions are detected either using spectroscopy or other means. Fractions containing the same protein can be pooled. After use, the column should be washed with two or three bed volumes of buffer. For long-term storage, the column should be washed with a volume of buffer containing 0.02% sodium azide.

## 7.8 AFFINITY CHROMATOGRAPHY

Affinity chromatography separates mixtures based on the principle of ligand-substrate interactions, such as antibody-antigen, enzyme-substrate, and receptor-agonist binding. These reversible interactions have a high affinity and allow for fairly specific separations. The first type of affinity chromatography developed involved insulin. Insulin was chemically bound to a stationary phase, and a mixture of proteins was loaded onto the column. Proteins that bind to insulin were retained on the column, whereas other proteins were eluted. After washing the column to remove unbound substances, the conditions were altered to reverse the interaction between the bound insulin and the bound proteins. Conditions to remove the bound protein include changes in pH, ionic strength, or passing a solution of the ligand (in the described example, insulin) through the column.

## 7.9 CAPILLARY ELECTROPHORESIS

Capillary electrophoresis was developed in the 1980s and combines the separation ability of electrophoresis (Chapter 5) with the speed and automation of high-performance liquid chromatography. Capillary electrophoresis is highly efficient and requires minute quantities of materials. The separated samples are eluted from the end of a capillary column and can be analyzed using quantitative detectors, such as mass spectrometers or UV-Vis detectors.

One of the downsides to slab electrophoresis is that in order to minimize heat damage, electrophoresis must be carried out at low voltages. This requires longer separation times. By carrying out electrophoretic separations in a capillary, higher voltages can be applied without adverse effects. This enables more rapid separations to occur.

In capillary electrophoresis (CE), samples are separated in a buffer-filled fused-silica capillary through the application of an electric field. Unlike HPLC, the solvent is not pumped through the capillaries in CE. The molecules are separated mainly by electroosmotic flow. Electroosmotic flow is the movement of a solvent caused by positively-charged ions migrating toward the negative electrode and carrying solvent molecules in the same direction. The surface of the capillary is negatively charged due to the intrinsic charge on the surface silanol groups. Cations from the buffer solution are attracted to the negatively-charged silanol groups. This creates an electrical double layer. When an electrical current is applied, the remaining cations in the buffer solution are attracted to the cathode (negatively charged electrode). Because the cations are solvated, they drag the bulk solvent with them, resulting in the overall flow of the solution toward the cathode away from the anode.

The movement of an ion in capillary electrophoresis is dependent on the charge and size of the ion as well as the viscosity of the separation medium.

$$\mu = \frac{q}{6\pi\eta r} \, ,$$

where    $\mu$ = mobility of the ion

           $q$ = net charge of the molecule

           $\eta$ = viscocity of the separation medium

           $r$ = radius of the ion

The velocity ($v$) of a particle or of the electroosmotic flow is determined by the product of the electric field strength (E) and the mobility.

$$v = \text{E}\mu$$

The rate of electroosmotic flow is generally greater than that of an individual particle. In the presence of electroosmosis, an ion's velocity will be the sum of its individual velocity and the velocity of the electroosmotic flow. Cations will have a total velocity that is greater than the velocity of the electroosmotic flow and will move faster through the capillary column. Neutral molecules will elute at the rate of the electroosmotic flow. Anions, whose individual velocities will be negative, will have a total velocity that is less than that of the electroosmotic flow and will elute last from the column. In this manner, substances can be separated using CE.

A number of different applications of capillary electrophoresis have been developed to improve the resolution of molecules. The different modes of separation will not be discussed in this manual. Interested readers should consult references about CE for more information.

## REFERENCES AND FURTHER READING


Barton, J. S., Tang, C.-F., and S. S. Reed. Dipeptide Sequence Determination: Analyzing Phenylthiohydantion Amino Acids by HPLC. *J. Chem. Ed.,* 77, 2000: 267.

Blumenfeld, F., and J. Gardner. Gel Filtration-An Innovative Separation Technique *J. Chem Ed.,* 62, 1985: 715-717.

Boyer, R. *Modern Experimental Biochemistry.* 3rd Edition. San Francisco, CA: Benjamin Cummings, 2000.

Bylkas, S. A., and L. A. Andersson. Microburger Biochemistry: Extraction and Spectral Characterization of Myoglobin from Hamburger. *J. Chem. Ed.,* 74, 1997: 426-430.

Clark, John M., and Robert L. Switzer. *Experimental Biochemistry.* 2nd Edition. New York, NY: W.H. Freeman and Company, 1977.

Copper, C. L. Capillary Electrophoresis Part I. Theoretical and Experimental Background. *J. Chem. Educ.,* 75, 1998: 343-347

Copper, C. L., and K. W. Whitaker. Capillary Electrophoresis: Part II. Applications. *J. Chem. Educ.,* 75, 1998: 347-351.

Gritter, R. J., Bobbitt, J. M., and A. E. Schwarting. *Introduction to Chromatography.* 2nd Edition. Oakland, CA: Holden-Day, Inc., 1985.

Makino, Y., et al. Profiling of Ectasy Tablets Seized in Japan. *Microgram Journal,* 1, 2003: 169-176.

Nelson, D. L., and M. M. Cox. *Lehninger Principles of Biochemistry.* 4th Edition. New York, NY: W.H. Freeman and Co., 2005.

Pugh, M. E., and E. Schultz. Assessment of the Purification of a Protein by Ion Exchange and Gel Permeation Chromatography. *Biochem. And Mol. Biol. Educ.,* 30, 2002: 179-183.

Rubinson, J. F., and Neyer-Hilvert. Integration of GC/MS Instrumentation into the Undergraduate Laboratory: Separation and Identification of Fatty Acids in Commercial Fats and Oils. *J. Chem. Ed.,* 74, 1997: 1106 – 1108.

Scott, R. P. W. *Techniques and Practices of Chromatography.* Marcel Dekker, 1995.

Skoog, D. A., Holler, F. J., and T. A. Nieman. *Principles of Instrumental Analysis.* 5th Edition. Orlando, FL: Harcourt Brace and Company, 1998.

### *Online References*

Applet for ion exchange chromatography
<http://www.biomoleculesalive.org/browse/
ViewResource.php?intResourceID=158&strType=Software&strTypeName=
Software+Resources&strTypeURL=http%3A%2F%2Fwww.biomoleculesalive.org
%2Fbrowse%2Fstep1_software.php&strDiscipline=fundamental>.

Online representation of GC in action
<http://ewr.cee.vt.edu/environmental/teach/smprimer/gc/gc-dg2.gif?>.

Other online applets for separations
<http://www.shsu.edu/%7Echm_tgc/sounds/sound.html>.

# Experiment 18. Sequence Determination of a Dipeptide

## Background

The sequence of a peptide or protein determines the structure, shape, and function of the molecule. It is therefore imperative to biochemists that the primary sequence of a peptide or protein be determined. Numerous methods for sequence determination exist. This laboratory involves creation of an N-terminal amino acid derivative, identification of the N-terminus of the peptide, and determination of the amino acid composition of a dipeptide. Using both pieces of information, the sequence of the dipeptide can be determined.

The compound phenylisothiocyanate (PITC) reacts with the amino terminus of the dipeptide and forms a phenylthiocarbamyl (PTC) derivative.

Phenylisothiocyanate (PITC)          Dipeptide          Phenylthiocarbamyl (PTC) Derivative

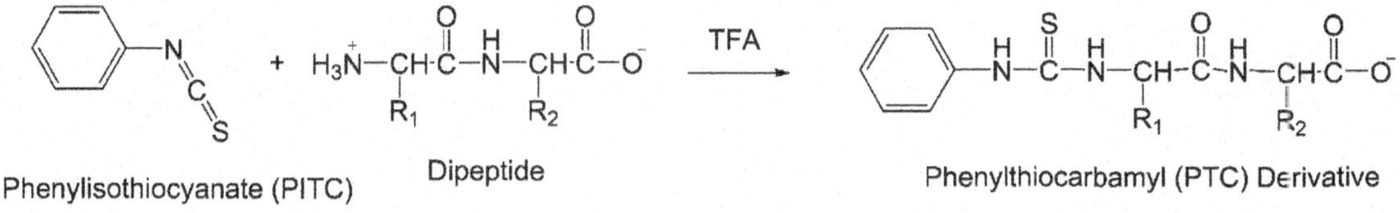

Hydrolysis of the peptide bond results in the formation of an anilinothiazolinone and the free C-terminal amino acid.

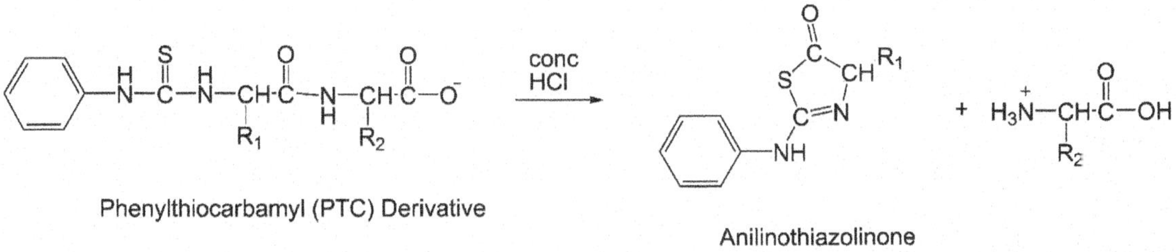

Phenylthiocarbamyl (PTC) Derivative          Anilinothiazolinone

Treatment with dilute acid converts the anilinothiazolinone to a phenylthiohydantoin (PTH) derivative.

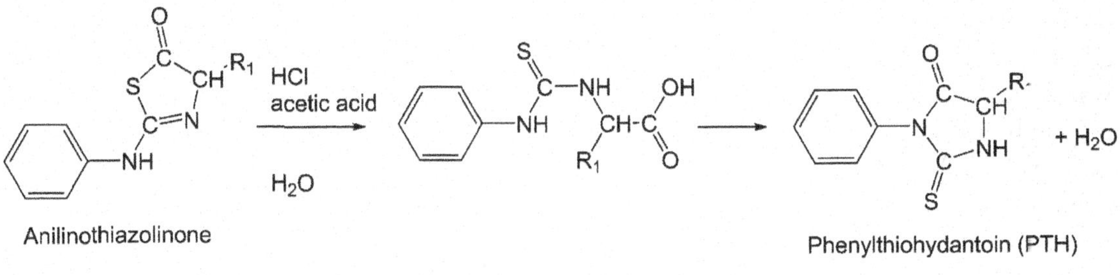

Anilinothiazolinone          Phenylthiohydantoin (PTH)

## Purpose of the Experiment

In this experiment, you will determine the sequence of an unknown dipeptide. This involves multiple steps in the laboratory, including hydrolysis of the dipeptide, determination of the amino acid composition, creation of an N-terminal phenylthiohydantoin amino acid derivative, and identification of the N-terminal PTH derivative and the free C-terminal amino acid.

## Prelaboratory Assignment

1) Do you expect the ninhydrin to stain N-terminal PTH derivatives in the same way as the free C-terminal amino acids? Explain your answer using appropriate chemical structures.

2) Predict the order of separation of amino acids on the paper chromatography support.

3) Predict the number of spots that you expect to observe in each fraction on the paper chromatography. Explain your predictions.

## Materials

Unknown dipeptides
Standard amino acid solutions
Ethanol
Triethylamine
Phenylisothiocyanate
Heptane
Ethyl acetate
Water
HCl
Glacial acetic acid
n-butanol
0.1% ninhydrin/5% collidine solution
6% nickel sulfate solution
Reaction vials
Conical flasks
Screwtop tubes
Water bath
Thermometer
Hydrolysis tube
Rotary evaporator
Oven
Watch glass
Bunsen burner
Chromatography jars
Cellulose TLC strips
Spray bottle
Forceps
Pipetters

## Safety

HCl, PITC, and TEA can be harmful if inhaled, swallowed, absorbed, or contacted with the skin. Wear gloves and dispense these chemicals in a hood.


## Experimental Methods

### *Preparation of Phenylthiohydrantoin Derivative*

1) Dissolve 2 mg of the dipeptide in 400 μl of 95% ethanol, 140 μl triethylamine (TEA), and 100 μl of phenylisothiocyanate (PITC) in a 1 mL vial.

2) Incubate the solution at 55°C for 30 minutes.

3) If the solution becomes cloudy, add a drop or two of 95% ethanol to clarify the solution.

4) Transfer the solution to a conical glass tube with a screw cap.

5) Add 500 μL of water and 1 mL of a 2:1 mixture of heptane-ethyl acetate. Shake well.

6) Remove the heptane-ethyl acetate layer with a Pasteur pipet. This layer should contain excess reagents and by-products.

7) Repeat Steps 5 and 6.

8) Transfer the aqueous phase to a watch glass. Evaporate the solvent in an oven at 55°C until dry. Alternatively, transfer the aqueous phase to a small conical flask, and dry it completely under vacuum with a rotary evaporator.

9) Cleave the PTC-dipeptide by adding 1 mL of 12 M HCl to the dried dipeptide. Heat at 55°C for 10 minutes.

10) Remove the HCl by evaporating on a watch glass in an oven or under vacuum.

11) Dissolve the dried sample in 500 μL of water. Extract the PTC-amino acid (N-terminal amino acid) residue twice with 500 μL of ethyl acetate. Save the aqueous layer for chromatography.

12) Combine the ethyl acetate fractions and evaporate to dryness on a watch glass in an oven or under vacuum.

13) To convert the dried derivative (from PTC-amino acid to PTH-amino acid), add 1 mL of a 2:1 (v/v) mixture of glacial acetic acid and 4.5 M HCl. Transfer the dissolved sample to a screw cap vial, and heat for 45 minutes at 55°C.

14) Evaporate the acid on a watch glass in an oven or under vacuum.

15) Dissolve the PTH-amino acid in 500 μL of water. Save the sample for chromatography.


### *Hydrolysis of the Dipeptide*

1) Dissolve 2 mg of dipeptide in 100 μL of 6 M HCl on a watch glass.

2) Draw the sample into a melting point capillary open on both ends. Carefully seal both ends with a Bunsen burner. Alternatively, transfer the dipeptide solution into a hydrolysis tube. Seal the open end of the tube with a Bunsen burner.

3) Place the sealed tube in an oven at 110°C for 12–15 hours (overnight).

4) Transfer the sample to a watch glass and evaporate to dryness in a 55°C oven. Alternatively, transfer the hydrolysate to a conical flask, and evaporate to dryness with a rotary evaporator.

5) Add 100 μL of water, and repeat the evaporation step.

6) Dissolve the sample in 100 μL of water. Save for chromatography.

### *Separation and Identification of Amino Acids Using Paper Chromatography*

1) Prepare developing solvent (n-butanol, acetic acid, and water in 12:3:5 v/v ratio).

2) Place a small amount of developing solvent into chromatography jars. Cover and let sit for at least 15 minutes to allow for saturation of the vapors in the jars.

3) Spot TLC plates with standard amino acids, total hydrolysate of the dipeptide, aqueous layer from derivatization, mixtures of standard amino acids, and PTH-derivative sample.

4) After the spots have dried, develop the TLC plates; remember to mark the solvent front at the end of developing the plate.

5) Allow plates to air-dry in the hood. Spray with the ninhydrin/collidine solution. Place plates in oven at 55°C until colored spots appear. Remove plates, circle any spots on the plates, and record any colors. Spray the plates with nickel sulfate solution and again record any colors.

6) Repeat the experiment so you can calculate $R_f$ values from at least two (preferably three) separate runs.

7) Calculate $R_f$ values for each sample. Determine the composition of your dipeptide.

## Optional Additional Explorations

Design an experiment to separate and analyze the protein lysate using HPLC instead of paper chromatography. Prior to any investigations, seek approval for the procedures from your instructor.

## Questions

1) What is the sequence of your unknown dipeptide sample?

2) Based on the results of your separation, rank the amino acids in order of increasing polarity (from lowest to highest polarity).

3) Did any of the amino acid(s) display unique properties during the laboratory? If so, based on the structure of the amino acid(s), explain why the amino acid(s) is (are) unique.

## Reference for Experiment

Barton, J. S., Tang, C.-F., and S. S. Reed. Dipeptide Sequence Determination: Analyzing Phenylthiohydantion Amino Acids by HPLC. *J. Chem. Ed.,* 77, 2000: 267.

# Experiment 19. Gel-Filtration Chromatography of Proteins

## Purpose of the Experiment

Gel filtration chromatography is frequently used to separate molecules based on size and shape. In addition to techniques such as SDS-PAGE, gel filtration is used to determine the molecular weights of compounds. The purpose of this experiment is to examine the separation of different proteins using gel filtration. Your instructor may ask that you use the same proteins from previous experiments (i.e., 13, 14, and 16) so that comparisons between purification methods can be made.

## Prelaboratory Assignment

Based on the information in Chapter 7 and other references, predict the order of elution of the proteins in the mixture used in the experiment.

## Materials

Sephadex G-75 columns
50 mM Tris-HCl (pH 8.0) buffer
A solution containing 3 or 4 known proteins from the table below (2 mg/mL of each
      protein) as directed by your instructor.

If you are running this experiment in conjunction with other experiments, use the same proteins in each experiment.

| Protein | Molecular Weight (Da) |
| --- | --- |
| Cytochrome c (horse) | 12,400 |
| α-lactalbumin (bovine) | 14,200 |
| Lysozyme (egg white) | 14,600 |
| Myoglobin (horse) | 17,600 |
| ß-lactoglobulin (bovine) | 18,400 |
| Trypsinogen (bovine) | 24,000 |
| Albumin (egg white) | 44,300 |
| Hemoglobin (bovine) | 64,500 |
| Bovine serum albumin | 66,000 |
| Glycogen phosphorylase b | 97,400 |
| ß-galactosidase | 116,250 |
| Myosin | 200,00 |

Unknown protein sample
Spectrophotometer
Electrophoresis apparatus
Precast Tris-HCl Polyacrylamide Gels, 12% separating gel, 4% stacking gel
If precast gels are not available, students must cast their own using
       30% acrylamide, 0.8% bis-acrylamide solution
       Separating gel buffer – 1.5 M Tris-Cl (pH 8.8)
       Stacking gel buffer – 0.5 M Tris (pH 6.8)
       10% ammonium persulfate
       TEMED
       Glass plates, spacers, casting apparatus
Electrophoresis buffer (15.1 g Tris, 72.0 glycine, water to 1L; do not adjust pH)
Sample buffer (7 mL 0.5 M Tris-Cl (pH 6.8), 3.0 mL glycerol, 1.2 mg
       bromophenol blue, water to 10 mL)
Protein marker mixture containing at least 6 different proteins (obtain information  about the
       composition from your instructor or the manufacturer)
Staining solution (1.0 g Coomassie Blue R-250, 450 mL methanol, 100 mL glacial acetic
       acid, water to 1 L)
Destaining solution (100 mL methanol, 100 mL glacial acetic acid, water to 1 L)


## Safety

Wear goggles. Use caution when using a centrifuge. Dispose of the solutions at the end of the
experiment in the proper manner, as indicated by the instructor.


## Experimental Methods

### *Gel Filtration of Proteins*

1)  Obtain a pre-poured and equilibrated Sephadex G-75 column (20-25 mL bed volume).

2)  Mix 400 μL of the mixture of known proteins with 100 μL of the unknown.

3)  Carefully pipet the mixed protein solution on top of the column.

4)  Once the protein sample has entered the gel, begin eluting the sample from the column
    by passing 50 mM Tris-HCl buffer through the column. Collect 1 mL fractions off of the
    column. In order to maintain a constant flow rate, keep a consistent volume of buffer
    above the top of the sephadex.

5)  If cuvettes require a volume greater than 1 mL, dilute 750 μL of the fractions with
    750 μL of water and measure the absorbance at 280 nm. If cuvettes hold a volume of
    1 mL or less, measure the $A_{280}$ of the sample without dilution. Make sure that you blank
    the spectrometer with 50 mM Tris-HCl buffer.

6)  Continue eluting the column, collecting fractions, and monitoring the absorbance until
    it appears that all of the proteins have eluted from the column. A plot of absorbance ver-
    sus fraction number will provide a series of peaks, similar to a chromatogram produced
    by a detector.

7) For the fractions containing proteins, mix 10 μL with 10 μL of SDS-PAGE buffer and resolve each fraction using SDS-PAGE. (See Experiment 14 for protocol.) Be sure to include a lane with markers and a lane with the original mixture (without the unknown protein).

8) Stain and destain the gel. Determine the identity of the protein in each fraction.

9) For the known proteins, create a plot of log of molecular weight versus elution volume. Using this plot, determine the MW of the unknown protein. You can also determine the MW of the unknown protein using the SDS-PAGE results.

## Optional Additional Explorations

Design experiments to investigate the fractionation of proteins via other gel matrixes with different-sized particles, such as G-25 or G-100. Compare the results of these experiments with those from the sucrose gradient. Prior to any investigations, seek approval for the procedures from your instructor.

## Questions

1) What was the actual order of elution in the experiment? Did it agree with your prediction? Explain.

2) Did the proteins elute in distinct zones? Was the separation using gel filtration adequate?

3) As demonstrated in Experiment 14, SDS-PAGE gels can be used to determine the molecular weights of proteins using the standard curve of log of molecular weight versus relative mobility. In this experiment, you used gel filtration, another method of molecular weight determination by creating a plot of log of MW versus elution volume. Is the curve useful? How do your results compare with those for SDS-PAGE? Which method for the determination of the molecular weight of proteins do you prefer?

4) Based on your findings, what is the identity of your unknown sample? Explain.

5) If you have conducted some of the other experiments in this text using the same proteins, compare your findings. Does one method provide information that differs from the others? Does one method separate the proteins better?

## References for Experiment

Blumenfeld, F., and J. Gardner. Gel Filtration-An Innovative Separation Technique *J. Chem Ed.*, 62, 1985: 715-717.

Pugh, M. E., and E. Schultz. Assessment of the Purification of a Protein by Ion Exchange and Gel Permeation Chromatography. *Biochem. And Mol. Biol. Educ.*, 30, 2002: 179-183.

# Experiment 20. Isolation and Characterization of Myoglobin from Hamburger

## Background

Myoglobin is a globular protein composed of a single polypeptide chain that transports oxygen to the tissues and stores the oxygen until the tissue requires it. Skeletal and cardiac muscles contain high concentrations of the protein, and the myoglobin gives these tissues their characteristic red color.

In a living cell or organism, the iron in the heme is in its $Fe^{2+}$ oxidation state, and the protein exists as either oxygenated (Oxy-Mb) or deoxygenated (deoxy-Mb). In the deoxy form, the iron exists in the $Fe^{3+}$ oxidation state and does not bind $O_2$. When Mb is isolated from an organism, the Oxy-Mb is slowly converted to Met-Mb. In this experiment, we will examine two stable forms of the protein found *in vitro*: Met-Mb [$Fe^{3+}$-$H_2O$-Mb] and Oxy-Mb [$Fe^{2+}$-$O_2$-Mb].

The two forms can be identified in the laboratory based on different spectral properties, as shown in Table 7.4.

Table 7.4 – Absorption Maxima for Myoglobin Complexes (reference Bylkas, S.A. 1997).

| Compound | Visible Region | | Soret Region (an intense absorption band for porphyrins in the blue region of the spectrum) | |
|---|---|---|---|---|
| Oxy-Mb | 635 nm | 504 nm | 409 nm ($\varepsilon$ = 179 mM$^{-1}$cm$^{-1}$) | |
| Met-Mb | 580 nm | 542 nm | 417 nm ($\varepsilon$ = 128 mM$^{-1}$cm$^{-1}$) | 348 nm |

One can interconvert the two *in vitro* forms of myoglobin using oxidation-reduction reactions. Treatment of Oxy-Mb with an oxidizing agent, such as potassium ferricyanide, converts the sample to Met-Mb.

$$Fe^{2+}\text{-}O_2\text{-Mb} + K_6[FeCN_6] \xrightarrow{-\ e^-} Fe^{3+}\text{-}H_2O\text{-Mb}$$

Alternatively, treatment of Met-Mb with a reducing agent, such as sodium dithionite, results in the formation of Oxy-Mb.

$$Fe^{3+}\text{-}H_2O\text{-Mb} + Na_2S_2O_4 \xrightarrow{+\ e^-} Fe^{2+}\text{-}O_2\text{-Mb}$$

## Purpose of the Experiment

In this experiment, myoglobin will be isolated from hamburger. The amount of Oxy-Mb and Met-Mb in the sample will be determined using spectral characterization of the myoglobin fractions and oxidation-reduction reactions. The crude myoglobin will be further separated using ion exchange and gel filtration chromatography. Bradford assay will be used to determine the protein concentrations of the fractions, and SDS-PAGE will be used to examine the purity of the samples obtained in the purification. A comparison of the purification methods will be made.

## Prelaboratory Assignment

1) In Step 11 of myoglobin extraction, you need to determine the total concentration of protein in the sample. Write a brief procedure for how you will determine the concentration of protein in your sample.

2) In the chromatography procedures, you will create a double Y axis plot. Explain how to create such a plot. What are you monitoring at the two different wavelengths (i.e., 280 and 417 nm)?

3) What is the difference in the two chromatography methods used? Based on your knowledge of myoglobin, which method would you predict to better isolate the myoglobin from hamburger?

## Materials

Centrifuge tubes
Glass rod with rubber policeman
Lean ground round steak (10 g)
20 mM potassium phosphate buffer (pH 5.6)
20 mM Tris (pH 7.5)
Spectrophotometer cuvettes
Pasteur pipets and bulbs
Graduated cylinder, 10 mL
0.1 M Potassium ferricyanide
0.1 M Sodium dithionite
Centrifuge
Spectrophotometer
Prepoured G-27 Sephadex column (30–40 mL bedvolume)
Prepoured carboxymethyl cellulose
(CMC) column (~10 mL bedvolume)
SDS-PAGE gel
Running buffer
Loading buffer
Staining solution
Destaining solution
1 mg/mL bovine serum albumin solution
Bradford reagent

## Safety

Wear goggles. Use caution when using a centrifuge. Dispose of the solutions at the end of the experiment in the proper manner, as indicated by the instructor.

## Experimental Methods

### *Myoglobin Extraction and Partial Purification*

1)  Obtain a 20 g sample of lean ground round steak. Place the sample into a centrifuge tube. Add 40 mL of 20 mM potassium phosphate buffer (pH 5.6).

2)  Carefully mix the sample with the glass rod for 1 minute to lyse the cells and release the myoglobin. Excessive or brutal mixing can result in release of fats and nucleic acids from the sample.

3)  The sample is centrifuged for 15 minutes at 20°C at 10,000 g. After centrifugation, there should be a white pellet, a brown supernatant containing Mb, and possibly a layer containing fat.

4)  Carefully filter the supernatant through glass wool to remove fat and unpelleted insoluble materials. Note the appearance of the filtrate.

5)  Pipet 2.0 mL of the supernatant into two cryogenic tubes, label with name and identification of the sample, and place in freezer.

6)  Place remaining supernatant in a disposable centrifuge tube. Label and place on ice.

7)  Dilute 1.0 mL of supernatant with 3.0 mL of potassium phosphate buffer.

8)  Following the instruction for the spectrophotometer, obtain a background scan from 700 to 300 nm using potassium phosphate buffer as a blank.

9)  Replace the buffer in the cuvette with the diluted supernatant sample from Step 7 above. Obtain a scan from 700 to 300 nm. Label all peaks on the spectrum, and print out a copy of the spectrum.

10) Determine the total mg of Mb in the sample using the $A_{417}$ data and the extinction coefficient of 7.57 mL•mg$^{-1}$cm$^{-1}$ or 128 mM$^{-1}$cm$^{-1}$.

11) Using procedure(s) in Experiment 5, determine the concentration of total protein in the sample. Determine the relative purity of the Mb by dividing the total mg of Mb by the total mg of protein.

### *Oxidation-Reduction Reactions and Gel Filtration Chromatography*

1)  Dilute 1 mL of original supernatant with 3 mL of buffer, and add 10 µL of 0.1 M potassium ferricyanide. Note any changes in the appearance of the sample.

2)  Blank the spectrophotometer from 700 to 300 nm with the phosphate buffer.

3) Obtain a scan of the extract with potassium ferricyanide from 700 to 300 nm. Label all peaks on the spectrum, and print out a copy of the spectrum.

4) Dilute 1 mL of original supernatant with 3 mL of buffer, and add 10 μL of 0.1 M sodium dithionite. Note any changes in the appearance of the sample.

5) Blank the spectrophotometer from 700 to 300 nm with the phosphate buffer.

6) Obtain a scan of the extract with sodium dithionite from 700 to 300 nm. Label all peaks on the spectrum, and print out a copy of the spectrum.

### *Further Purification of Myglobin Using Sephadex G-75*

1) Carefully load 1.0–1.5 mL of the original supernatant onto the top of a pre-equilibrated Sephadex G-75 column. Try not to disturb the top of the column. Allow the sample to enter the column by gravity.

2) After the sample has entered the gel, carefully pour potassium phosphate buffer onto the column to a buffer height of 2.0 cm above the column matrix. While maintaining the constant buffer height at the top of the column (in order to maintain a constant flow rate), collect 1.0 mL fractions. While collecting the fractions, determine the flow rate for the column.

3) Continue to collect fractions until all of the red color moving through the column has been eluted. Collect several fractions after all of the red color has eluted.

4) Add 1 mL of buffer to each of the fractions, and determine the absorbance of each at 417 and 280 nm. (Use phosphate buffer as a blank.)

5) Determine the total mg of Mb, total protein in the sample, and relative purity of the Mb in each fraction using procedure in Steps 10-11, Part I.

6) Freeze the fractions containing Mb for characterization.

7) Create a double Y axis plot of $A_{417}$ and $A_{280}$ versus fraction number. A double Y plot is one in which two layers are superimposed in one graph. The X axis for both layers is the same, while the Y axes are independent. One y-axis is placed on the left side of the plot, while the other is on the right.

### *Further Purification of Myoglobin Using Carboxymethylcellulose (CMC)*

1) Carefully load 1.0–1.5 mL of the original supernatant onto the top of a pre-equilibrated CMC column. Try not to disturb the top of the column. Allow the sample to enter the column by gravity.

2) After the sample has entered the gel, carefully pour one column volume (~10 mL) of 20 mM potassium phosphate buffer (pH 5.6) onto the column. Collect 2 mL fractions from the column.

3) After all of the phosphate buffer has entered the column, add 20 mM Tris buffer (pH 7.5) to the column, and collect 2 mL fractions. Continue to collect fractions until all of the red color moving through the column has been eluted.

4) Add 2.0 mL of distilled water to each fraction collected after the Tris buffer was added to the column. Determine the absorbance of each solution at 417 and 280 nm, including the ones collected with the phosphate buffer as the eluent.

5) Determine the total mg of Mb, total protein in the sample, and relative purity of the Mb using procedure in Steps 10-11, Part I.

6) Freeze the fractions containing Mb for characterization.

7) Create a double Y plot of $A_{417}$ and $A_{280}$ versus fraction number.

### *Characterization of the Fractions*

1) Thaw reserved fractions.

2) Determine the composition of each fraction using SDS-PAGE. See Experiment 14 for experimental details. Using protein concentrations, calculate the volume of each sample to be loaded on the gel so it contains 20–40 µg of protein (without exceeding a volume of 20 µL). If 20–40 µg cannot be obtained in less than 20 µL, then 20 µL of the sample will be used. If 20-40 µg of protein occurs in a volume less than 20 µL, dilute the appropriate volume with enough water to create a 20 µL sample.

## Questions

1) Using Beer's law, calculate the concentration of Oxy-Mb and Met-Mb in your sample. Show your calculations.

2) Based on the protein concentration data, approximately what percentage of total protein in the hamburger is myoglobin?

3) Think about the color of freshly isolated ground beef and ground beef that has been in the refrigerator for several days. Based on your findings in this experiment, what inferences can you make about the color of the meat?

4) What form of iron, $Fe^{2+}$ or $Fe^{3+}$, is more readily absorbed by the body? What do your results suggest about the "dietary" availability of iron from "aged" meat?

5) Which purification method, Sephadex G-75 or CMC, was a better separations method?

## References for the Experiment

Bylkas, S. A., and L. A. Andersson. Microburger Biochemistry: Extraction and Spectral Characterization of Myoglobin from Hamburger. *J. Chem. Educ.,* 74, 1997: 426-430.

Marcoline, A. T., and T. E. Elgren. A Thermodynamic Study of Azide Binding to Myoglobin. *J. Chem. Educ.,* 75, 1998: 1622.

Pugh, M. E., and E. Schultz. Assessment of the Purification of a Protein by Ion Exchange and Gel Permeation Chromatography. *Biochem. And Mol. Biol. Educ.,* 30, 2002: 179-183.

# Experiment 21. Isolation of Tyrosinase from Mushrooms

## Purpose of the Experiment

In Experiment 11, the kinetic activity of tyrosinase was investigated. Details about the enzyme can be found in Chapter 4. The enzyme can be isolated from various fruits and vegetables. A crude extract from mushrooms will be partially purified using ammonium sulfate precipitation and column chromatography. The extent of purification will be explored.

## Prelaboratory Assignment

1) In the characterization of fractions portion of the experiment, you need to determine the total concentration of protein in the sample. Write a brief procedure for how you will determine the concentration of protein in your sample.

2) What is the difference in the two chromatography methods used? Why is it necessary to use two types of chromatography following the ammonium sulfate precipitation?

## Materials

Mushrooms
100 mM phosphate buffer (pH 7.3)
Cheesecloth
Centrifuge
Ammonium sulfate
Sephadex G-75 column (30–40 mL bedvolume)
Q-Sepharose column (10 mL bedvolume)
100 mM phosphate buffer (pH 7.3) containing 150 mM NaCl
Test tubes
Spectrophotometer
Cuvettes
SDS-PAGE gel
Running buffer
Loading buffer
Staining solution
Destaining solution
1 mg/mL bovine serum albumin solution
Reagent for protein concentration determination
L-DOPA (5 mM in 0.1 M phosphate buffer, pH 7); prepare fresh – do not chill, as
        precipitation may occur
M phosphate buffer at pH 7

## Safety

Wear goggles. Use caution when using a centrifuge. Dispose of the solutions at the end of the experiment in the proper manner, as indicated by the instructor.

## Experimental Procedure

### *Crude Extraction of Tyrosinase*

1) Briefly blend mushrooms in blender with 100 mM phosphate buffer, pH 7.3 (use approximately 40 mL of buffer for 20 g of mushroom).

2) Filter the extract through a double layer of cheesecloth.

3) Centrifuge the sample at 10,000 g for 15 minutes.

4) Reserve 1 mL of sample for future characterization; store on ice.

### *Ammonium Sulfate Isolation*

1) To a portion of the crude extract, add ammonium sulfate to 30%. (Refer to Chapter 6 and Table 6.1.) Slowly add ammonium sulfate to the tube. Do not add all of the salt at the same time. Gently shake the tube after each addition until the salt dissolves. Keep the tube on ice during the addition.

2) Incubate the tube on ice for 5 minutes after all of the salt has dissolved.

3) Centrifuge the tube at 10,000 xg for 10 minutes. Be sure that a counterbalance tube is used in the centrifugation.

4) Separate the pellet from the supernatant. Reserve 1 mL of the supernatant on ice.

5) Resuspend pellet in 5 mL of 100 mM phosphate buffer. Reserve the sample on ice.

6) To remaining supernatant, add enough ammonium sulfate to reach 60%. Do not add all of the salt at the same time. Gently shake the tube after each addition until the salt dissolves. Keep the tube on ice during the addition.

7) Incubate the tube on ice for 5 minutes after all of the salt has dissolved.

8) Centrifuge the tube at 10,000 xg for 10 minutes. Be sure that a counterbalance tube is used in the centrifugation.

9) Separate the pellet from the supernatant. Reserve 1 mL of the supernatant.

10) Resuspend pellet in 5 mL of 100 mM phosphate buffer. Reserve.

11) Measure $A_{280}$ of all reserved samples.

12) For fractions containing protein, mix 50 µL of the fraction with 500 µL of 5 mM DOPA. Observe the fractions after 5 minutes. The fractions that change color should be retained as containing tyrosinase.

13) Complete the following with fractions containing enzyme. Place 950 µL of 5 mM DOPA in two cuvettes. Blank the instrument at 475 nm. Add 50 µL of a fraction containing the enzyme to the sample cuvette. Mix well and rapidly. Measure the change in absorbance every 15 seconds for 4 minutes. For this assay, one unit of tyrosinase is defined as the amount of enzyme that changes one absorbance unit per minute.

14) Repeat Step 13 with all other fractions containing enzyme.

15) Freeze all fractions that contain tyrosinase for future characterization.

### Gel Filtration Chromatography

1) Carefully load 1.0–1.5 mL of the original extract onto the top of a pre-equilibrated Sephadex G-75 column. Try not to disturb the top of the column. Allow the sample to enter the column by gravity.

2) After the sample has entered the gel, carefully pour 100 mM potassium phosphate buffer (pH 7.3) onto the column to a buffer height of 2.0 cm above the column matrix. While maintaining the constant buffer height at the top of the column (in order to maintain a constant flow rate), collect 1.0 mL fractions. While collecting the fractions, determine the flow rate for the column.

3) Continue to collect fractions; look for color. Measure $A_{280}$ of samples and look for peaks.

4) For fractions containing protein, mix 50 μL of the fraction with 500 μL of 5 mM DOPA. Observe the fractions after 5 minutes. The fractions that change color should be retained as containing tyrosinase.

5) Complete the following with fractions containing enzyme. Place 950 μL of 5 mM DOPA in two cuvettes. Blank the instrument at 475 nm. Add 50 μL of a fraction containing the enzyme to the sample cuvette. Mix well and rapidly. Measure the change in absorbance every 15 seconds for 4 minutes. For this assay, one unit of tyrosinase is defined as the amount of enzyme that changes one absorbance unit per minute.

6) Repeat Step 5 with all fractions containing tyrosinase.

7) Freeze the fractions for additional characterization.

8) Create a double Y axis plot of units of tyrosinase and $A_{280}$ versus fraction number. A double Y plot is one in which two layers are superimposed in one graph. The X axis for both layers is the same, while the Y axes are independent. One y-axis is placed on the left side of the plot, while the other is on the right.

### Further Purification Using Q-Sepharose

1) Carefully load 1.0–1.5 mL of the original extract onto the top of a pre-equilibrated Q-Sepharose in 100 mM phosphate (pH 7.3). Try not to disturb the top of the column. Allow the sample to enter the column by gravity.

2) After the sample has entered the gel, carefully pour one column volume (~10 mL) of 100 mM potassium phosphate buffer (pH 7.3) onto the column. Collect 2 mL fractions from the column.

3) After all of the phosphate buffer has entered the column, add 100 mM phosphate phosphate buffer (pH 7.3) containing 150 mM NaCl to the column, and collect 2 mL fractions. Continue to collect fractions; look for color. Measure $A_{280}$ of samples and look for peaks.

4) For fractions containing protein, mix 50 μL of the fraction with 500 μL of 5 mM DOPA. Observe the fractions after 5 minutes. The fractions that change color should be retained as containing tyrosinase.

5) Complete the following with fractions containing enzyme. Place 950 μL of 5 mM DOPA in two cuvettes. Blank the instrument at 475 nm. Add 50 μL of a fraction containing the enzyme to the sample cuvette. Mix well and rapidly. Measure the change in absorbance every 15 seconds for 4 minutes. For this assay, one unit of tyrosinase is defined as the amount of enzyme that changes one absorbance unit per minute.

6) Repeat Step 5 with all fractions containing tyrosinase.

7) Freeze the fractions for additional characterization.

8) Create a double Y axis plot of units of tyrosinase and $A_{280}$ versus fraction number. A double Y plot is one in which two layers are superimposed in one graph. The X axis for both layers is the same, while the Y axes are independent. One y-axis is placed on the left side of the plot, while the other is on the right.

### *Characterization of the Fractions*

1) Thaw reserved fractions.

2) Using an appropriate assay from Experiment 5, determine the concentration of protein in each of the reserved fractions.

3) Determine the composition of each fraction using SDS-PAGE. See Experiment 14 for experimental details. Using protein concentrations, calculate the volume of each sample to be loaded on the gel so that it contains 20–40 μg of protein (without exceeding a volume of 20 μL). If 20–40 μg cannot be obtained in less than 20 μL, then 20 μL of the sample will be used. If 20–40 μg of protein occurs in a volume less than 20 μL, dilute the appropriate volume with enough water to create a 20 μL sample.

## Optional Additional Explorations

Conduct kinetic analysis on the isolated tyrosinase. Design experiments to purify tyrosinase from other species of mushroom or other fruits or vegetables and compare their kinetic activity. Prior to any investigations, seek approval for the procedures from your instructor.

## Questions

1) What is the specific activity (total units of enzymatic activity/mg of total protein) for each fraction?

2) Tyrosinase is involved in the browning of fruits and vegetables that occurs during injury. Do you think that the amount of tyrosinase in a bruised mushroom would be greater than the amount of the enzyme in an uninjured mushroom?

3) Which purification method, ammonium sulfate precipitation, Sephadex G-75, or Q-Sepharose, was a better separations method?

## References for the Experiment

Gilly, R., et al. Resveratrol and a Novel Tyrosinase in Carignan Grape Juice. *J. Agric. Food Chem.*, 49, 2001: 1479-1485.

Okot-Kotber, M., et al. Activation of Polyphenol Oxidase in Extracts of Bran from Several Wheat (*Tritcum aestivum*) Cultivars Using Organic Solvents, Detergents, and Chaotropes. *J. Agric. Food Chem.*, 50, 2002: 2410-2417.

Zhang, X., et al. Characterization of Tyrosinase from the Cap Flesh of Portabella Mushrooms. *J. Agric. Food Chem.*, 47, 1999: 374-378.

# Experiment 22. Separation and Identification of Fatty Acids in Commercial Oils

## Background

Lipids are a diverse class of biological molecules that can be extracted from a cell using organic solvents. The fatty acid content of commercial oils can be determined using gas chromatography. The fatty acids are isolated from the triacylglycerols by saponification.

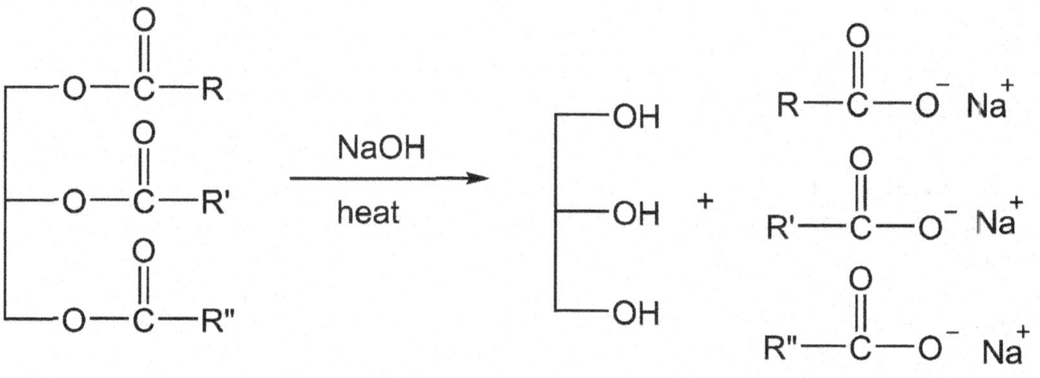

Fatty acid methyl esters (FAMEs) are synthesized by treating the samples with $BF_3$ in methanol.

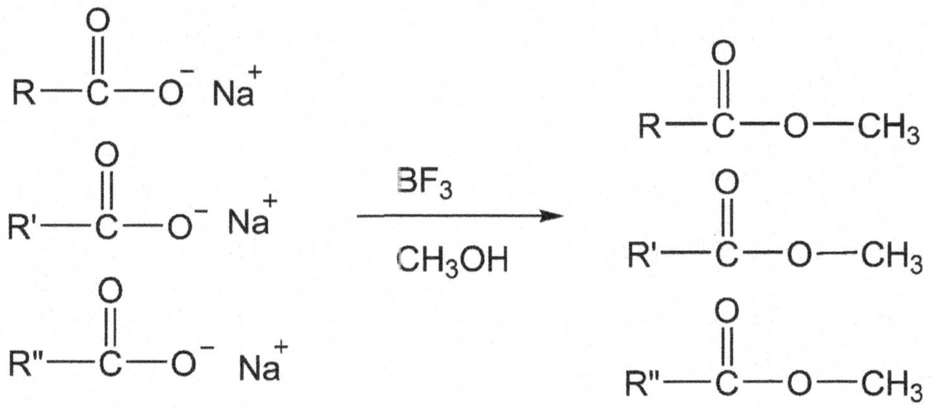

GC/MS analysis can be used to simultaneously identify and quantitate the FAMEs.

## Purpose of the Experiment

The purpose of this experiment is to identify and determine the percentages of fatty acids in commercial oils.

## Prelaboratory Assignment

1) Using chemical literature, determine the fatty acid compositions of the three oils to be analyzed in this experiment.

2)  What is chemically occurring during each step of the isolation and synthesis of fatty acid methyl esters? What is the role of each reagent used?

3)  Pick one of the oil compositions in your answer to Question 1, and predict the order of elution from the GC. Explain your answer.


## Materials

Corn oil
Peanut oil
Safflower oil
0.5 M NaOH dissolved in methanol (methanolic NaOH)
Anhydrous $MgSO_4$
NaCl
12% $BF_3$ in methanol
Standard fatty acid methyl esters (palmitic, oleic, linoleic, stearic)
Methanol
Hexane
NaOH


## Safety

Wear goggles. Dispose of the solutions at the end of the experiment in the proper manner, as indicated by the instructor.


## Experimental Methods

### *Isolation of the Fatty Acids and Synthesis of Fatty Acid Methyl Esters*

1)  Weigh 50-75 mg sample of oil into a 18 x 150 mm test tube.

2)  Add 3 mL of 0.5 M methanolic NaOH.

3)  Heat sample at 60°C until a homogenous solution is obtained.

4)  Add 5 mL of $BF_3$/methanol.

5)  Boil sample for 2 to 3 minutes.

6)  After cooling the sample, transfer the solution into a separatory funnel containing 25 mL of hexane and 20 mL of saturated NaCl solution.

7)  Shake well, but gently. Allow layers to separate. If an emulsion forms, add a small amount of water to the mixture to help break up the emulsion. Allow the layers to separate. Drain off the aqueous layer.

8)  Pour the hexane layer into a dry flask. The hexane layer is dried with ~ 1 g of anhydrous $MgSO_4$. Decant the hexane layer into a clean, dry container.

9) Cap the solution of fatty acid methyl esters to prevent air oxidation, and store in the refrigerator until it is analyzed by GC.

### *GC/MS Analysis of Standard Fatty Acid Methyl Esters and Commercial Oil Samples*

1) Familiarize yourself with the GC/MS instrument. Table 7.5 lists the GC and MS conditions to use in this experiment.

Table 7.5 – GC and MS Conditions.

| Column | |
|---|---|
| Carrier | 99% He, flow rate |
| Injector temperature | 250°C |
| Detector temperature | 280°C |
| Column temperature program | 50-260°C @ 10°C/min; hold 5 minutes at 260°C |
| Solvent delay on MS | 3.5 minutes |
| Mass range | 50–500 |

2) Inject a sample of each of the known fatty acid methyl esters (FAMEs) into the GC/MS instrument, and run the method described in Table 7.5. Obtain total ion chromatographs for each sample.

3) Create a mixture containing known amounts of each of the four standard FAMEs. Inject a sample of the mixture into the GC/MS; obtain a total ion chromatograph and a summary of peak area percentages for the sample.

4) Calculate the response factor for each methyl ester using the following formula:

$$\text{Response factor} = \frac{\text{peak area, given standard}}{\text{peak area, palmitic acid standard}} \times \frac{\text{concentration, palmitic acid standard}}{\text{concentration, given standard}}$$

Small unidentified peaks will be assumed to have a response factor of 1.00.

5) Inject a sample of each of the FAMEs extracted from the oils into the GC/MS; obtain total ion chromatographs and a summary of peak area percentages for each sample.

6) Using the response factors for each standard, the mass percentage of each of the FAMEs is calculated using the following equations:

a) Correction of peak area of FAME for varying detector sensitivity

$$\text{Corrected area} = \frac{\text{peak area of given compound}}{\text{response factor for that compound}}$$

b)   Percentage of each FAME in the sample

$$\text{Percentage} = \frac{\text{corrected area for given component}}{\text{total of corrected areas}}$$

## Questions

1)  Compare the experimental percent compositions for each of the oils with literature values. If variations from the literature values exist, offer some reason(s) for the difference.

2)  Why are the FAMEs rather than free fatty acids used for the gas chromatography?

3)  Using the data from your experiment, write the order of gas chromatography elution (first to last) for the following sets of fatty acid methyl esters. Explain your answer.

a)   14:0, 20:0, 12:0, 16:0, 18:0
b)   16:0, 18:0, 16:1$^{\Delta 9}$, 18:1$^{\Delta 9}$

# CHAPTER 8

# DNA Processing Enzymes and DNA Manipulation

DNA manipulation and cloning have become an increasingly important means of determining molecular information about cellular architecture and genetic makeup. The Human Genome project and other programs studying the genetic codes of other organisms are providing information about diseases and hereditary. The fields of genomics and proteomics are offering details about cellular compositions that are assisting forensics, medicine, and agriculture. Since DNA-based technologies are increasingly growing, this chapter will only focus on a few topics within this complex field.

## 8.1 DNA RESTRICTION ENZYMES

Restriction enzymes are proteins that cleave DNA at specific sites within the nucleic acids and are referred to as restriction endonucleases, because of this targeting of the interior sequence. The enzymes are isolated from bacteria and are often harvested for use by scientists. Restriction endonucleases were discovered in the early 1960s and function in bacteria to cleave foreign DNA from other organisms, such as viruses. Bacteria methylate their native DNA with methylase to distinguish the host DNA from the invading DNA.

There are three types of restriction endonucleases. Each enzyme recognizes a specific sequence within the DNA. Type I and III are large, multisubunit complexes that contain both endonuclease and methylase activities and require ATP hydrolysis for activity. Type I enzymes cleave at random sites that can be found more than 1000 base pairs from the recognition sites. Type III endonucleases cleave sequences about 25 base pairs from the recognition sequences. Type II restriction enzymes were first isolated in 1970 by Hamilton Smith. These enzymes do not require ATP, are smaller in size, and cleave the DNA within the recognition sequence. Type II enzymes are frequently used in DNA cloning, recombination, and analysis. Daniel Nathans first demonstrated the use of Type II restriction enzymes for mapping and analysis of genes. An example of this application is Restriction Fragment Length Polymorphisms, described in Box 5.1.

The enzymes are named using the system suggested by Smith and Nathans, in which the restriction endonuclease is named for the bacteria from which they are isolated. For example, the first enzyme that was isolated from *Providencia stuartii* was named *Pst*I. The first letter of the genus, P, is followed by the first two letters of the specific epithet, st, and Roman numerals are used to identify the number of the enzyme isolated. *Pst*I is the first enzyme isolated, and *Pst*II is the second enzyme isolated. Some of the names also indicate the strand of bacterium from which the enzyme was isolated. For example, *Bam*HI is the first enzyme isolated from *Bacillus amyloliquefaciens*, strain H.

Each enzyme recognizes and cuts a specific sequence of nucleotides. Most of the recognized sites are 4 to 8 base pairs (bp) in length and are palindromes, sequences with twofold symmetry that read the same in the forward direction as in the reverse direction. Table 8.1 lists the restriction sites of numerous restriction enzymes. Recall the name of the enzyme is related to the organism from which it is isolated and does not provide information about the restriction sequence. The list in Table 8.1 is not all inclusive, and sequence information is available in a number of references, including textbooks and catalogs. The Website www.rebase.neb.com contains a restriction enzyme database (REBASE) that lists all known restriction enzymes, their commercial availability, buffer conditions, and literature references.

Table 8.1 – Example Type II Restriction Enzymes and Their Target Restriction Sites.

| Enzyme | Target Sequence | Enzyme | Target Sequence |
|---|---|---|---|
| *Alu*I | 5'------ A G C T------3'<br>3'------ T C G A------5' | *Bam*HI | 5'------ G G A T C C------3'<br>3'------ C C T A G G------5' |
| *Cla*I | 5'------ A T C G A T------3'<br>3'------ T A G C T A------5' | *Eco*RI | 5'------ G A A T C------3'<br>3'------ C T T A A G------5' |
| *Hae*III | 5'------ G G C C------3'<br>3'------ C C G G------5' | *Hin*dIII | 5'------ A A G C T T------3'<br>3'------ T T C G A A------5' |
| *Hpa*II | 5'------ C C G G------3'<br>3'------ G G C C------5' | *Pst*I | 5'------ C T G C A G------3'<br>3'------ G A C G T C------5' |

Each restriction endonuclease has a unique cleavage site. Two different enzymes may recognize the same sequence but cleave at a different place. In Table 8.1, the site of the cleavage is indicated with the arrows and falls into one of three different categories: "blunt" ends, 5' overhangs, or 3' overhangs. Restriction enzymes such as *Alu*I cut straight across the DNA in the middle of its sequence creating "flush" or "blunt" ends. (See Figure 8.1.) Other enzymes cut the DNA in an offset fashion with either a 5' or 3' overhang, such as *Bam*HI and *Pst*I, respectively. (See Figures 8.2 and 8.3.) These overhanging single-strand DNA segments are often referred to as "sticky" or "cohesive" ends because they can bind to complementary sticky ends through hydrogen bonding. The DNA strand breaks can be resealed using an enzyme called DNA ligase.

```
5'------ A G C T------ 3'      AluI        5'------ A G-3'  +  5'-C T ------ 3'
3'------ T C G A------ 5'     ------->     3'------ T C-5'     3'-G A ------ 5'
```

Figure 8.1 – Example of restriction enzyme that cleaves DNA with blunt ends.

```
                                            5'------ G
5'------ G G A T C C------3'     BamHI       3'------ C C T A G------ 5'
3'------ C C T A G G------5'    ------->
                                                        +

                                                  5'------ G A T C C------3'
                                                          G------5'
```

Figure 8.2 – Example of restriction enzyme that cleaves to produce a sticky end with a 5' overhang.

```
                                            5'------ C T G C A------ 3'
                                            3'------ G

5'------ C T G C A G----- 3'     PstI                   +
3'------ G A C G T C----- 5'    ------->

                                                          G------ 3'
                                            3'------ A C G T C----- 5'
```

Figure 8.3 – Example of restriction enzyme that cleaves to produce a sticky end with a 3' overhang.

One unit of enzyme is defined as the quantity of enzyme that is required to digest 1 µg of standard DNA in one hour under optimal temperature and buffer conditions. The standard DNA that is frequently utilized is Lambda (λ) DNA. The bacteriophase λ infects *E. Coli* and is one of the most studied phages. Its DNA has been isolated, is 48,502 base pairs in length, and contains restriction sites for many endonucleases. Table 8.2 shows the seven restriction sites for *Hind* III on λ DNA.

Table 8.2 – Restriction Sites on λ DNA for *Hind* III.

| Cut site from zero point (basepairs) | Fragment generated from zero or the last cut (kilobases) |
|---|---|
| 23130 | 23.1 |
| 25157 | 2.03 |
| 27479 | 2.32 |
| 36895 | 9.42 |
| 37459 | 0.564 |
| 37584 | 0.125 |
| 44141 | 6.56 |
| 48502 (not a cleavage site, but the end of the piece of DNA) | 4.36 |

Each restriction enzyme has its own optimal reaction buffer. DNA samples and solutions used for digestions should be free of contaminants, such as phenol, chloroform, alcohol, salt, detergents, and chelating agents. Trace amounts of these substances will inhibit or inactivate the restriction endonucleases. The enzymes are often commercially provided in a buffered solution containing 50% glycerol. The reaction conditions should contain 10% glycerol or less in order to prevent sensitivity of the enzymes to high concentrations of the glycerol. Most of the restriction enzymes are very heat labile and are stored at -20°C. During use, the enzymes should be thawed briefly on ice to prevent denaturation. Most enzymes react at 37°C, but a few have alternative optimal temperatures; for example, *Taq*I and *Sma*I react at 65°C and 25°C, respectively. You should check the manufacturer's specifications for buffer composition and optimal reaction temperature.

Once the DNA is cleaved, the resulting fragments can be separated by gel electrophoresis or HPLC. (See Chapters 5 and 7.) Frequently, the separations are used to visualize and characterize the number of fragments, such as in restriction fragment length polymorphism (RFLP) or characterization of a DNA segment. Also, scientists will isolate a specific fragment for future studies, such as creation of a recombinant vector. (See Figure 8.4.)

The number of times that a restriction enzyme will cleave a piece of DNA depends on the sequence of the nucleic acid and the length. Generally, longer pieces of DNA will have a greater chance of repeating the restriction site than smaller pieces of DNA. The probability of repeating the sequence also depends on the length of the recognition sequence. For DNA in which the four nucleotides are present in random sequence but equal abundance, a 4 bp recognition sequence will occur in the overall sequence once every $4^4$ or 256 bp. A 5 bp sequence will occur in this DNA sequence every $4^5$ or 1024 bp. Genomic DNA is very large and is cleaved multiple times by specific restriction endonucleases. This occurs at a lower rate than in the random sequence DNA because the sequences in genomic DNA are not random and the nucleotides are not found in equal abundances. Also, large DNA fragments are sensitive to shearing during handling and will break into pieces of approximately 50,000 to 100,000 bp.

Plasmid DNA is a circular piece of DNA in bacteria that replicates separately from the chromosomal DNA. It is generally used in cloning experiments to express foreign genes in bacteria in order to make copies of the gene product, protein. In order to insert the piece of DNA to be expressed, restriction enzymes are frequently used to cut the plasmid DNA and the target DNA. After the two are cut with the restriction enzymes, the resulting pieces of DNA are joined together to form a new circular plasmid, often referred to as a recombinant vector. (See Figure 8.4.)

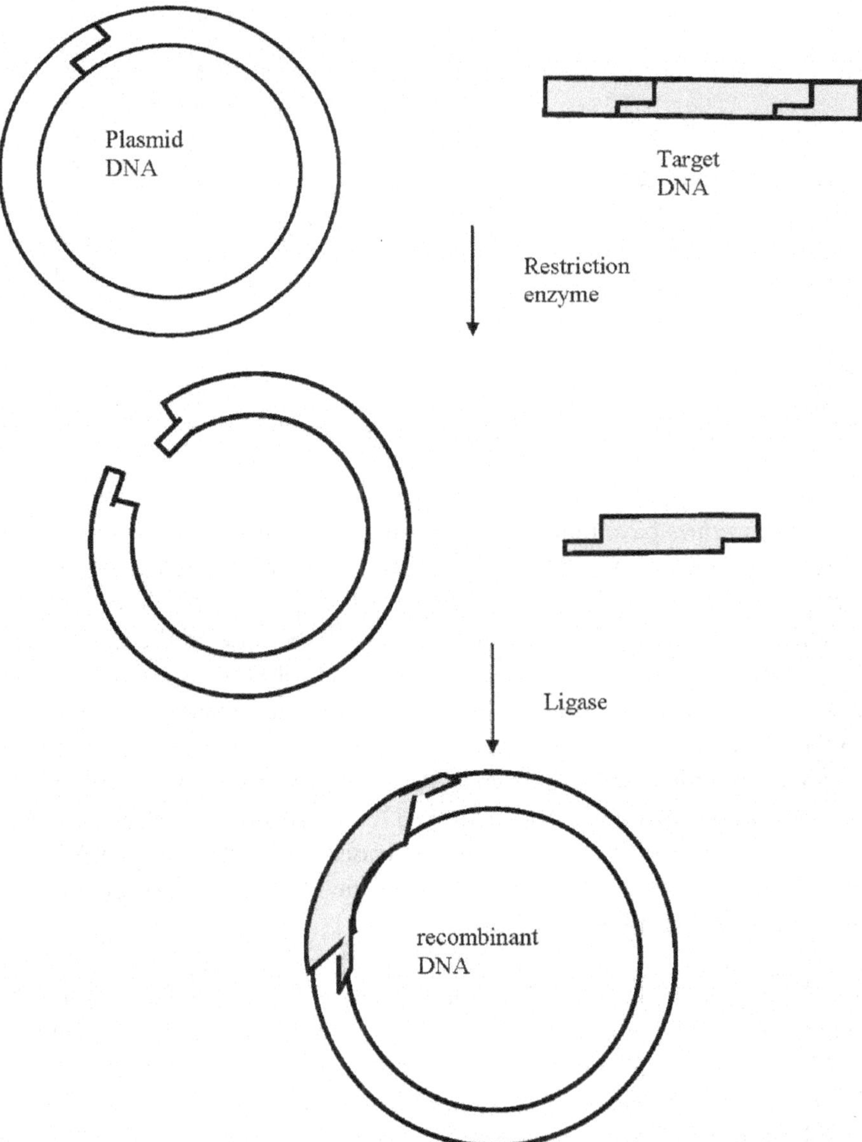


Figure 8.4 - Creation of a recombinant vector.

It is important that the cleavage sites on a piece of DNA are known so that the DNA is cleaved at an appropriate place rather than in a random location. As mentioned before, restriction enzymes are also often used to generate fragment patterns that can be used to identify DNA for forensic or medical purposes. A restriction map is a drawing of a piece of DNA indicating the locations of the restriction sites for specific enzymes. If the DNA is circular and a restriction enzyme has only one site, the result of the cleavage reaction will be a piece of linear DNA. If the DNA is linear and contains a single restriction site, the resulting fragmentation is two fragments, the sizes of which depend on the location of the restriction sites. Figure 8.5 shows an example of a restriction map for λ DNA. As seen in the figure and indicated in Table 8.2, *Hind*III cleaves the lambda DNA at seven sites, resulting in eight fragments that can be observed by electrophoresis. *Eco*RI, on the other hand, cleaves the lambda DNA at only five sites.

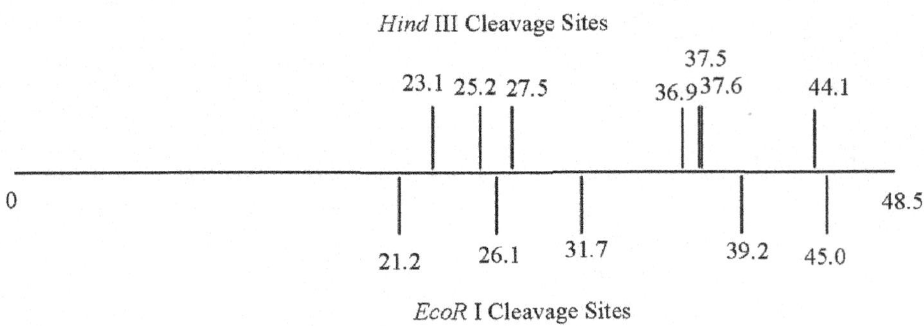


Figure 8.5 – Restriction map for lamba DNA.

Just as restriction enzymes have unique cleavage sites, individual DNA sequences have different patterns of restriction enzyme sites. Due to the existence of alleles, alternative forms of genes, each individual has a unique DNA sequence. Examples of alleles include blood type and eye color. Another reason that individual DNA sequences differ is because of the degeneracy within the genetic code. In transcription, a copy of one of the strands of DNA is created as a piece of RNA. The RNA can be used to create protein sequences. A three-letter codon (sequence) of RNA indicates the correct amino acid to be added to the sequence. The degeneracy in the code is that more than one codon often indicates the same amino acid. For example, UUU and UUC both encode for phenylalanine. In this manner, a change in the DNA sequence can often occur without changing the resulting protein sequence. This is another reason why individual DNA has different patterns of restriction enzyme sites. These variations in the DNA sequences are observed in restriction fragment length polymorphism (RFLP).

In RFLP, DNA samples are treated with restriction enzymes, and the resulting fragments are separated using agarose gel electrophoresis. If the DNA used in the analysis is small in size, such as plasmid DNA, the discrete banding patterns can be easily observed after gel electrophoresis. However, larger pieces of DNA, such as human chromosomes, result in multiple bands that are often not resolved, resulting in a smear. Scientists use Southern blotting (see Chapter 5) with a specific probe to visualize differences in the banding patterns. The patterns derived during the restriction enzyme cleavage are analyzed, and the results can be used to differentiate species and/or individuals (i.e., different people, strains, etc.) from one another. This technique has become an important tool for solving crimes, identifying remains after terrorist attacks, determining genetically engineered foods, etc.

## 8.2 DNA TOPOISOMERASES

DNA supercoiling is a way to compact the DNA into structures such as chromosomes. The term "supercoiling" describes the coiling of a coil and can be simply observed in a telephone cord. On a traditional land line phone, a telephone cord is a coiled wire. If this coil is twisted onto itself, supercoiling occurs. A similar phenomenon occurs with the DNA double helix. Positive supercoiling of DNA occurs when the right-handed double helix is twisted in a right-handed fashion. The result is overwinding of the DNA, which leads to distortion of the structure. Negative supercoiling occurs when the double helix is twisted in a left-handed fashion, and the DNA is underwound and distorted. Supercoiling (in both forms) frequently occurs when the DNA strands are separated in DNA replication and transcription. The cell must have enzymes to help deal with this phenomenon.

As discussed in Chapter 5, molecules are electrophoretically separated based on the applied electric field, charge, and size and shape. Assuming a DNA sample is uniform (i.e., same charge) and the separation is conducted under the same conditions, DNA topoisomers should separate based on shape. Supercoiled DNA is more compact and will migrate faster through an agarose gel, whereas relaxed DNA is larger and will be retarded. Linear DNA (as well as different topoisomers) migrates intermediately between supercoiled and relaxed. Figure 8.6 demonstrates the migration of different topoisomers in an agarose gel.

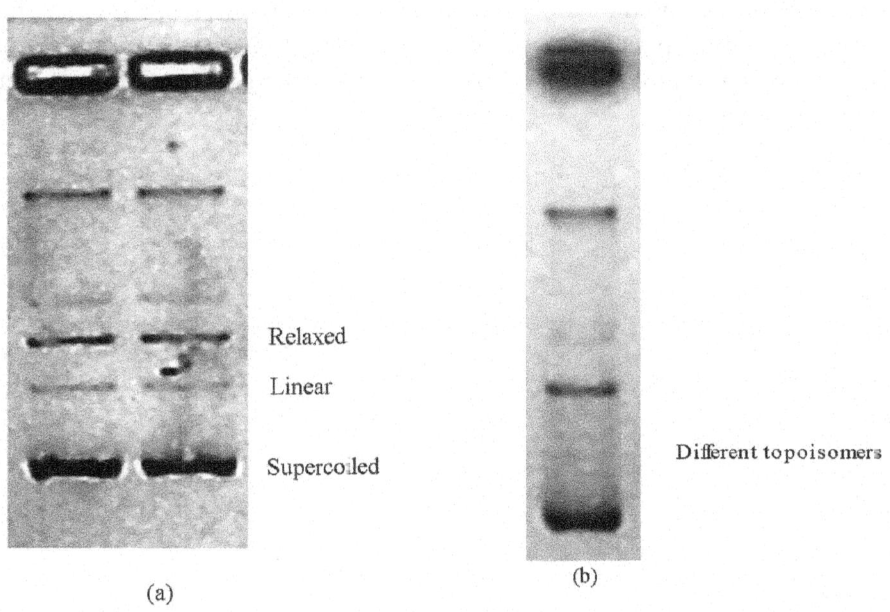


Figure 8.6 – Gel electrophoresis of different DNA topoisomers.

Topoisomerases are essential enzymes that regulate the topology of DNA. These enzymes have been isolated from prokaryotes, eukaryotes, and viruses. Topoisomerases are classified by the way in which they alter the DNA topology. Type I enzymes introduce a transient nick into one of the DNA strands and allow the structure to untwist. The nick is then resealed. Type II enzymes cleave both DNA strands and pass an intact strand through the break. After resealing the break, the topology has changed. Eukaryotic topoisomerase II relaxes supercoiled DNA or decatenates DNA. Catenated DNA occurs when two circular pieces of DNA are interlinked, similar to the metal circles used by magicians or the trademark Olympic circles. In decatenation, a double-stranded break is introduced in one of the DNA pieces, and the intact strand is passed through the break, resulting in separation of the DNA strands. Prokaryotic topoisomerase II is also called gyrase. This enzyme can relax and decatenate DNA but can also introduce supercoiling and catenation.

Both classes of enzymes cleave the DNA by forming a covalent intermediate between a tyrosine in the enzyme and a phosphate group within the DNA backbone. Topoisomerase II requires ATP hydrolysis in order to pass the intact strand through the break, whereas topoisomerase I does not require ATP for its mechanism. The topoisomerase II mechanism also requires a divalent cation, such as $Mg^{2+}$.

# 8.3 INHIBITORS OF DNA TOPOISOMERASES

A number of pharmaceutical agents used to treat cancer and bacterial and parasitic infections target topoisomerases. These drugs act by either enhancing the cleavage reaction or preventing the resealing of the DNA breaks. In the presence of these agents, DNA strand breaks result, and, if unresolved, these breaks can result in cell death. Due to the essential role in DNA replication and transcription, topoisomerases are present in rapidly replicating cells, such as cancer, bacteria, or parasites.

Leading anticancer agents include the topoisomerase I inhibitor camptothecin and its derivatives and topoisomerase II inhibitors like epipodophyllotoxins, anthracyclines, anthracenediones, amino-acridines, and ellipticines. DNA gyrase is the target of several classes of antibiotic drugs, including quinolones and coumarins. Figure 8.7 shows the structure of a number of topoisomerase poisons used as pharmaceutical agents.

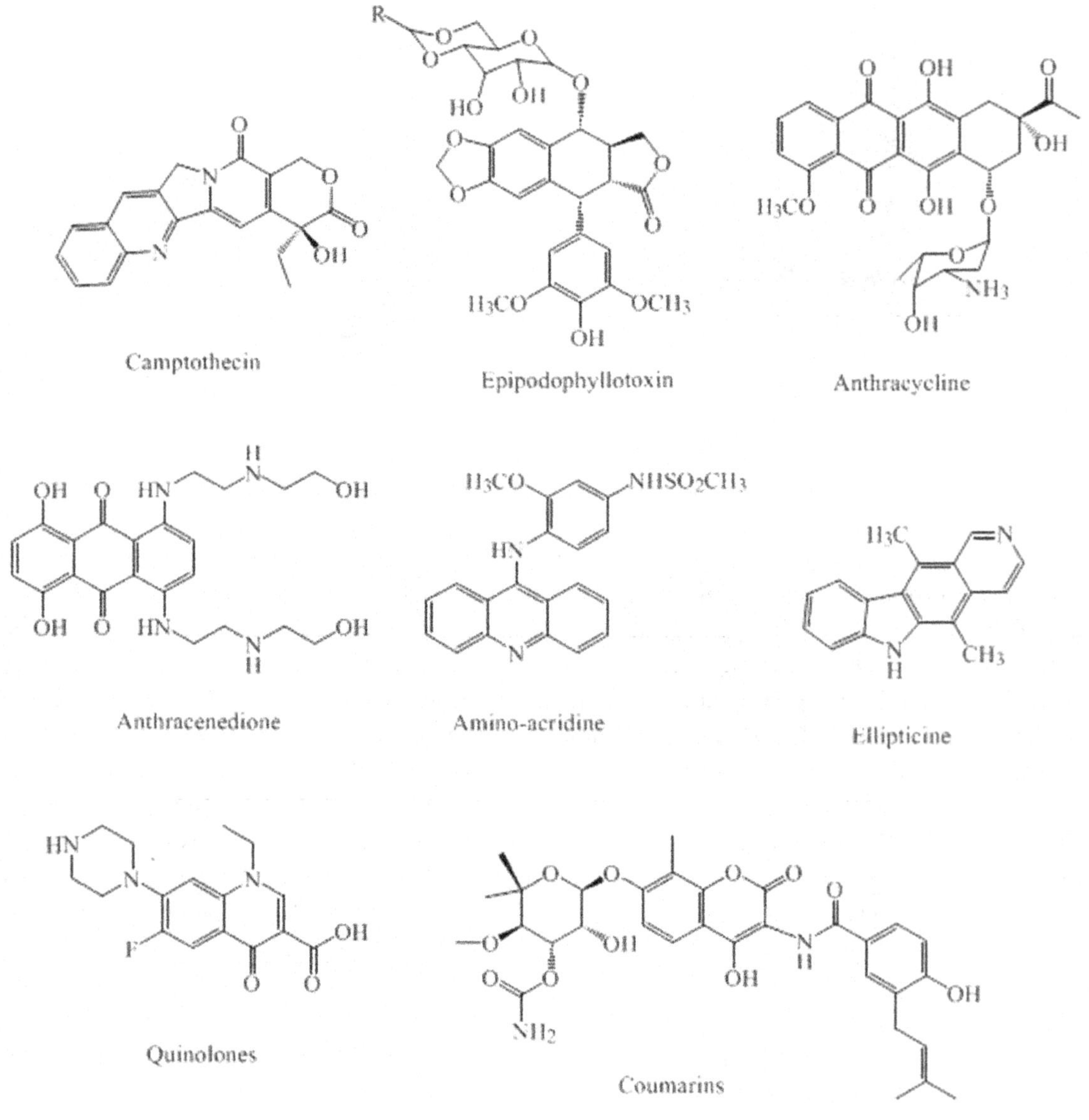

Figure 8.7 – Structures of topoisomerase poisons.

## 8.4 POLYMERASE CHAIN REACTION

DNA cloning and characterization often requires creating large quantities of DNA from smaller concentrations. Forensic and anthropological studies provide miniscule amounts of nucleic acids, and, while a scientist would like to identify and characterize the DNA, the small DNA concentration is not conducive to in-depth study. The conception of the Polymerase Chain Reaction (PCR) by Kary Mullis in 1983 provided the means to amplify copies of DNA starting with these very small quantities. Dr. Mullis shared the Nobel Prize in Chemistry in 1983 for his invention of PCR method.

PCR can be used to synthesize large quantities of DNA from small quantities of template as long as the flanking sequences around the target sequence are known. Primers—small, single-stranded DNA sequences that are complementary to the ends of the DNA region to be amplified—are designed and utilized in the PCR reaction. A typical PCR mixture contains the DNA template, primers, buffer with $Mg^{2+}$ or another suitable divalent cation, four deoxyribonucleotides, and a thermostable DNA polymerase, such as *Taq* polymerase. Primer number 1 targets the beginning of the DNA sequence to be amplified; the second primer is complementary to the other end of the target region, but on the opposite strand.

The procedure involves a number of cycles consisting of three steps, as illustrated in Figure 8.8. The first step, denaturation, involves heating the sample to approximately 94°C, which results in the separation of the DNA strands. The denaturation temperature depends on the composition of the DNA (see Section 3.8) and is determined experimentally. Once the strands are separated, the sample is cooled so the primers in the mixture can bind (anneal) to the complementary sequences in the target DNA. Again the annealing temperature is dependent on the composition of DNA and is often around 50°C. The third step, extension, involves heating the sample to 72°C so that the polymerase can synthesize the DNA starting with the two primers. The cycle is repeated 35–40 times, using an automated thermocycler. In this manner, the DNA is exponentially amplified, $2^n$ times, where n is the number of cycles. After the first cycle, two DNA strands exist. After the second cycle, four copies of DNA are present in the sample. After 30 cycles, $2^{30}$, or over one million copies, of the DNA exist. In this manner, DNA is rapidly copied.

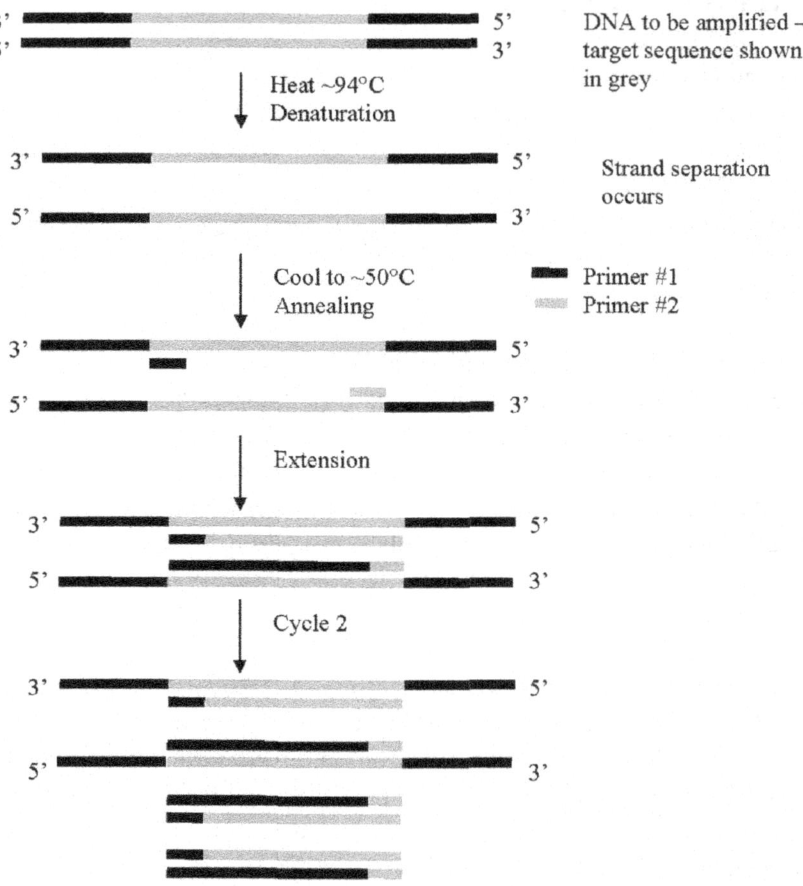


Figure 8.8 - Three steps in a cycle of the polymerase chain reaction.

The key to PCR is the identification and use of a thermostable DNA polymerase. *Taq* polymerase was originally isolated from the thermophyllic bacterium *Thermus aquaticus*. The high temperature utilized in the denaturation step would destroy the activity of most enzymes (see Section 4.3), but this is not observed with thermostable polymerases. For routine PCR reactions, 0.5–2.5 units of *Taq* are used in a 25-50 µL reaction. The commercial preparation of the enzyme differs in terms of yield, length, and fidelity. It is therefore important that the reaction conditions for PCR be optimized each time a new batch of *Taq* is obtained.

For greater fidelity or for target DNAs that are longer than a few thousand bases, other DNA polymerases or cocktails of two or more polymerases are frequently employed. Bio-Rad sells a cocktail DyNAzyme that combines the high proofreading functionality of *Tbr* with the high efficiency of *Taq*. Combinations of *Taq* and *Pfu* are available as *TaqPlus* Large PCR System (Strategene) and Expand Long Template PCR system (Roche Applied Science).

The primers are synthetic oligonucleotides that are used to initiate DNA synthesis. The primers must be carefully designed to obtain high yields of the DNA products, to prevent amplification of unwanted sequences, and to enable further possible manipulation of the product. As a general rule, longer primers are more specific and prevent aberrant amplifications. However, if the sequence is too long, it will be very expensive to synthesize or obtain commercially. The recommended length of a primer is 18–25 nucleotides. As mentioned before, two primers are utilized in the PCR reaction. The members of this primer pair must have similar physical properties (i.e., $T_m$ and length) to enable the annealing to occur at the same time in the reaction but must have different sequences to prevent complementary binding of the primers together. The primers must also lack self-complementary sequences that could cause hairpins or other secondary structures to form within the primer itself, thus preventing binding to the DNA target.

In Section 8.1, RFLP analysis was mentioned, and the practical uses of this technique were briefly listed. In order for RFLP to be conducted, an ample supply of DNA must be isolated from the sample to be analyzed. This isolation is time-consuming and often produces only small quantities of DNA sample. Scientists have now combined PCR with RFLP in order to speed the process of isolating sufficient quantities of DNA. The DNA from a sample (such as hair, blood, or bone) is amplified using PCR to the levels required for RFLP analysis. The DNA is then digested with restriction enzymes, and the fragments are separated using gel electrophoresis. PCR/RFLP allows scientists to analyze more samples in a shorter period of time.

## REFERENCES AND FURTHER READING


Boyer, R. *Modern Experimental Biochemistry.* 3rd Edition. San Francisco, CA: Benjamin Cummings, 2000.

Farrell, S. O., and R. T. Ranallo. *Experiments in Biochemistry. A Hands-on Approach.* Florence, KY: Brooks Cole Thomas Learning, 2000.

Keck, M. V. DNA Topology Analysis in the Undergraduate Biochemistry Laboratory. *J. Chem. Educ.,* 77, 2000: 1471-1473.

Kima, P. E., and M.E. Rasche. Sex Determination Using PCR. *Biochem. Mol. Biol. Educ.,* 32, 2004: 115-119.

MacAfee, L. K. *DNA Topology Studies Using Atomic Force Microscopy.* Honors Thesis, University of Mary Washington, 2005.

Nelson, D. L., and M. M. Cox. *Lehninger Principles of Biochemistry.* 4th Edition. New York, NY: W.H. Freeman and Co., 2005.

Sanbrook, J., and D. W. Russell. *Molecular Cloning. A Laboratory Manual.* 3rd Edition. Cold Spring Harbor, NY: Cold Spring Harbor Laboratory Press, 2001.

Tweedie, J. W., and K. M. Stowell. Quantification of DNA by Agarose Gel Electrophoresis and Analysis of the Topoisomers of Plasmid and M13 DNA Following Treatment with a Restriction Endonuclease or DNA Topoisomerase I. *Biochem. And Mol. Biol. Educ.,* 33, 2005: 28-33.

Wang, J. C. DNA Topoisomerases: Why So Many? *J. Biol. Chem.,* 266, 1991: 6659-6662.

*Websites*

Bio-Rad Laboratories. <http://www.bio-rad.com>.

Fermentas Life Science. <http://www.fermentas.com>.

Maxim Biotech. <http://www.maximbio.com/>.

New England Biolabs. <http://www.neb.com>.

Roberts, R. J., and D. Macelis. "REBASE. The Restriction Enzyme Database."
     <http://rebase.neb.com >.

Roche Applied Science. <http://www.roche-applied-science.com/>.

Stratagene. <http://www.stratagene.com/homepage/>.

# Experiment 23. Analysis of DNA Restriction Fragments

## Purpose of the Experiment

In this experiment, DNA samples will be cleaved with several restriction enzymes, and the resulting fragments will be separated using gel electrophoresis. The discrete banding patterns can be used to identify an individual DNA sample, as in a forensic analysis, or can be used to generate a restriction enzyme map. In the mock forensic laboratory exercise, plasmid DNA is utilized to provide discrete banding patterns rather than the smearing that could result from using human DNA. This is designed to avoid using Southern blotting.

## Prelaboratory Assignment

1) If a circular piece of DNA has three sites for a particular restriction enzyme, into how many fragments will that restriction enzyme cut the DNA?

2) If a linear piece of DNA has three sites for a particular restriction enzyme, into how many fragments will that restriction enzyme cut the DNA?

3) Construct a restriction map of a linear fragment of DNA using the following data. Your map should indicate the relative positions of the restriction sites, along with distances from the ends of the molecule to the restriction sites and between restriction sites:

| DNA | | Sizes of Fragments (bp) |
|---|---|---|
| i. | uncut DNA | 10,000 |
| ii. | DNA cut with *Eco*RI | 8000, 2000 |
| iii. | DNA cut with *Bam*HI | 5000, 5000 |
| iv. | DNA cut with *Eco*RI + *Bam*HI | 5000, 3000, 2000 |

## Materials

DNA samples (isolated from suspects and from crime scene)
λ DNA
Reaction buffer for endonucleases
*Pst*I
*Eco*RI
*Hind*III
Unknown restriction endonuclease
Water bath
Electrophoresis apparatus
Agarose
1X TBE buffer (89 mM Tris base, 89 mM boric acid, and 2 mM EDTA, pH 8.0)
DNA electrophoresis loading buffer (30% glycerol, 0.25% (w/v) bromophenol blue, 0.25% (w/v) xylene cyanol)
*Hind* III λ DNA fragments
Ethidium bromide or alternative stain

## Safety

Ethidium bromide is a suspected carcinogen. Wear gloves when handling ethidium bromide solutions. Wear UV goggles to protect your eyes when using the UV transilluminator. Dispose of the solutions at the end of the experiment in the proper manner, as indicated by the instructor.

## Experimental Methods

### I.    *Mock Forensic Analysis Using RFLP*

In this experiment, you will be provided with a sample of DNA that was isolated from a crime scene and with the DNA isolated from some suspects. Using RFLP analysis, identify the perpetrator of the crime.

1) Obtain DNA samples from your instructor. The DNA provided is in a buffered solution that is appropriate for the restriction enzymes tested. If you have three suspects and one sample from the crime scene, you will be preparing a total of 16 tubes.

Table 8.3 – Samples for Mock Forensic Analysis.

| Reaction Component | Tube 1 | 2 | 3 | 4 |
|---|---|---|---|---|
| Water | 2 µL | 1 µL | 1 µL | --- |
| DNA | 8 µL | 8 µL | 8 µL | 8 µL |
| EcoRI | --- | 1 µL | ---- | 1 µL |
| PstI | ---- | ---- | 1 µL | 1 µL |

2) Centrifuge the tubes in a microcentrifuge for a few seconds in order to draw all of the liquid to the bottom of the tube.

3) Incubate the tubes at 37°C for 20-30 minutes.

4) Meanwhile, pour a 1% agarose gel in TBE buffer. Based on the manufacturer's information, determine the volume of agarose solution required to create the gel. Measure the appropriate amount of agarose, and place it in an Erlenmeyer flask with the determined volume of TBE buffer. Gently heat the solution over a hotplate, in a microwave, or in an autoclave until the agarose melts and a homogenous solution is created. Avoid boiling or charing the agarose.

5) Cool the solution to 55°C, and pour into the gel cast. Remove any bubbles and insert the comb.

6) Allow the gel to harden. Once hardened, remove the gel comb and place the gel into the electrophoresis chamber. Cover the gel with TBE buffer.

7) After incubation, mix the DNA samples with 2 µL of loading buffer. If necessary, mix the *Hind*III λ DNA fragments with 0.2 volumes of loading buffer (some manufacturers prepare DNA ladder samples containing loading buffer).

8) Slowly load the samples into the wells in the agarose gel using a micropipetter. Load the DNA standards (ladder) on both the right and left sides of the gel.

9) Close the electrophoresis apparatus. Apply a voltage of 1–10 V/cm until the bromophenol blue has migrated within 1 cm of the end of the gel.

10) Turn off the power, disconnect the leads, and remove the gel from the electrophoresis chamber.

11) Stain the gel by soaking in TBE buffer containing 0.5 µg/mL ethidium bromide for 10 to 30 minutes.

12) Visualize the DNA bands by illuminating the gel with UV light using a transiluminator.

13) Document the gel using photography.

14) Determine the size of the DNA restriction fragments. Based on the RFLP fragments, identify the criminal.


## II.    *Creation of a Restriction Map for an Unknown Restriction Endonuclease*

In this experiment, you will be provided with a sample of lambda DNA, *EcoR* I, *Hind* III, and an unknown restriction enzyme. Using the data obtained, a restriction map for the unknown enzyme will be created.

1) Prepare the mixtures listed in Table 8.4 in microcentrifuge tubes on ice.

Table 8.4 – Samples for Restriction Mapping.

| Reaction Component | Tube 1 | 2 | 3 | 4 | 5 | 6 | 7 |
|---|---|---|---|---|---|---|---|
| 10 x reaction buffer | 1 µL | 1 µL | 1 µL | 1 µL | 1 µL | 1 µL | 1 µL |
| Water | 6 µL | 6 µL | 6 µL | 5 µL | 5 µL | 5 µL | 4 µL |
| lambda DNA | 2 µL | 2 µL | 2 µL | 2 µL | 2 µL | 2 µL | 2 µL |
| *EcoR* I | 1 µL | ---- | ---- | 1 µL | 1 µL | ---- | 1 µL |
| *Hind*III | ---- | 1 µL | ---- | 1 µL | ---- | 1 µL | 1 µL |
| Unknown endonuclease | ---- | ---- | 1 µL | --- | 1 µL | 1 µL | 1 µL |

2) Centrifuge the tubes in a microcentrifuge for a few seconds in order to draw all of the liquid to the bottom of the tube.

3) Incubate the tubes at 37°C for 20-30 minutes.

4) Meanwhile, pour a 1% agarose gel in TBE buffer. Based on manufacturer's information, determine the volume of agarose solution required to create the gel. Measure the appropriate amount of agarose and place in an Erlenmeyer flask with the determined volume of TBE buffer. Gently heat the solution over a hotplate, in a microwave, or in an autoclave until the agarose melts and a homogenous solution is created. Avoid boiling or charing the agarose.

5) Cool the solution to 55°C, and pour into the gel cast. Remove any bubbles and insert the comb.

6) Allow the gel to harden. Once hardened, remove the gel comb and place the gel into the electrophoresis chamber. Cover the gel with TBE buffer.

7) After incubation, mix the DNA samples with 2 µL of loading buffer. If necessary, mix the *Hind*III λ DNA fragments with 0.2 volumes of loading buffer (some manufacturers prepare DNA ladder samples containing loading buffer).

8) Slowly load the samples into the wells in the agarose gel using a micropipetter. Load the DNA standards (ladder) on both the right and left sides of the gel.

9) Close the electrophoresis apparatus. Apply a voltage of 1–10 V/cm until the bromophenol blue has migrated within 1 cm of the end of the gel.

10) Turn off the power, disconnect the leads, and remove the gel from the electrophoresis chamber.

11) Stain the gel by soaking in TBE buffer containing 0.5 µg/mL ethidium bromide for 10 to 30 minutes.

12) Visualize the DNA bands by illuminating the gel with UV light using a transilluminator.

13) Document the gel using photography.

14) Determine the size of the DNA restriction fragments.


## Data Analysis

1) Plot the log of DNA fragment (base pairs) versus the migration distance for the DNA standards (lambda DNA *Hind*III fragments; see Table 8.2). Draw a smooth curve that fits the data. Using the migration distance for the fragments in each sample, determine the size of the DNA fragments for each restriction endonuclease.

2) Using the fragmentation from the single and multiple digestions, create a restriction map to the best of your ability for the unknown restriction endonuclease.

## Questions

1) Based on the data obtained, which suspect is implicated in the crime? How definite is your evidence? Do you think it would be enough to convict?

2) How many cleavage sites exist for your unknown endonuclease? Are there regions of the lambda DNA that are cleaved by the unknown endonuclease and not by *Eco*RI or *Hind*III?

## References for Experiment

Bio-Rad Laboratories, http://www.bio-rad.com [accessed July 2005].

Farrell, S. O., and R. T. Ranallo. *Experiments in Biochemistry. A Hands-on Approach.* Florence, KY: Brooks Cole Thomas Learning, 2000.

# Experiment 24. Visualization of DNA Topological Changes

## Purpose of the Experiment

In this experiment, DNA samples will be treated with topoisomerase enzymes, and the changes in topology will be observed using gel electrophoresis.

## Prelaboratory Assignment

1) Explain the role of sodium dodecyl sulfate (SDS) and proteinase K in the reactions conducted in the laboratory.

2) Predict the outcome (draw the predicted gel electrophoresis) of the reaction of topoisomerase I with supercoiled plasmid.

## Materials

DNA samples, relaxed and supercoiled plasmids
10X Reaction buffer for topoisomerase I (10 mM Tris-Cl, pH 7.5, 100 mM NaCl or KCl, 1 mM
      PMSF and 1.0 mM mercaptoethanol)
10X Reaction buffer for gyrase (350 mM Tris-HCl, pH 7.5, 250 mM KCl, 40 mM $MgCl_2$,
      20 mM dithiothreitol, 18 mM spermidine, 10 mM ATP, 65% (w/v) glycerol and
      1 mg/mL bovine serum albumin)
Topoisomerase I
Gyrase
Sodium dodecyl sulfate (SDS), 10% w/v
Proteinase K (10 mg/mL)
Water bath
Electrophoresis apparatus
Agarose
1X TBE buffer (89 mM Tris base, 89 mM boric acid, and 2 mM EDTA, pH 8.0)
DNA electrophoresis loading buffer (30% glycerol, 0.25% (w/v) bromophenol blue, 0.25% (w/v)
      xylene cyanol)
Ethidium bromide or alternative DNA stain

## Safety

Ethidium bromide is a suspected carcinogen. Wear gloves when handling ethidium bromide solutions. Wear goggles to protect your eyes when using the UV transilluminator. Dispose of the solutions at the end of the experiment in the proper manner, as indicated by the instructor.

## Experimental Methods

### *I. Effect of Topoisomerases on Different Topological Forms of DNA*

1) Prepare two sets of the mixtures listed in Table 8.5 in microcentrifuge tubes on ice. One set of tubes should use the reaction buffer for topoisomerase I, and the other set should use the reaction buffer for gyrase.

2) Centrifuge the tubes in a microcentrifuge for a few seconds in order to draw all of the liquid to the bottom of the tube.

3) Add 2 μL of the appropriate enzyme (either topoisomerase I or gyrase) to tubes 3 and 4.

Table 8.5 – Samples for Topoisomerase Studies.

| Reaction Component | Tube | | | |
|---|---|---|---|---|
| | 1 | 2 | 3 | 4 |
| 10 x reaction buffer | 2 μL | 2 μL | 2 μL | 2 μL |
| Water | 16 μL | 16 μL | 14 μL | 14 μL |
| Supercoiled DNA | 2 μL | ——— | 2 μL | ——— |
| Relaxed DNA | ——— | 2 μL | ——— | 2 μL |

4) Incubate the tubes at 37°C for 30 minutes.

5) Meanwhile, pour a 1% agarose gel in TBE buffer. Based on manufacturer's information, determine the volume of agarose solution required to create the gel. Measure the appropriate amount of agarose, and place in an Erlenmeyer flask with the determined volume of TBE buffer. Gently heat the solution over a hotplate, in a microwave, or in an autoclave until the agarose melts and a homogenous solution is created. Avoid boiling or charing the agarose.

6) Cool the solution to 55°C and pour into the gel cast. Remove any bubbles and insert the comb.

7) Allow the gel to harden. Once hardened, remove the gel comb and place the gel into the electrophoresis chamber. Cover the gel with TBE buffer.

8) After incubation, stop the reactions by adding 2 μL of 10% SDS, 2 μL of proteinase K solution, and 4 μL of loading dye. Incubate the samples at 37°C for 15-20 minutes.

9) Slowly load the samples into the wells in the agarose gel using a micropipetter.

10) Close the electrophoresis apparatus. Apply a voltage of 1–10 V/cm until the bromophenol blue has migrated within 1 cm of the end of the gel.

11) Turn off the power, disconnect the leads, and remove the gel from the electrophoresis chamber.

12) Stain the gel by soaking in TBE buffer containing 0.5 μg/mL ethidium bromide for 10 to 30 minutes.

13) Visualize the DNA bands by illuminating the gel with UV light using a transilluminator.

14) Document the gel using photography.

### II. Time Study of the Topoisomerase Reactions

Determine which form of DNA is acted upon by each enzyme, and conduct the same experiment as in Part I with that form of DNA, but incubate the enzyme with the DNA for varying times from 0 to 30 minutes. The best way to ensure consistent results is to begin the reaction at varying times such that all samples are stopped *at the same time* through the addition of SDS/proteinase K (Step 8) at the same time. For example, a time point of 25 minutes is started by adding the enzyme to its reaction tube, 5 minutes *after* some enzyme has been added to another tube to start the reaction for 30 minutes.

## Optional Additional Explorations

Design experiments to investigate the effect of cofactors (i.e., divalent cations, ATP) on the reactions catalyzed by topoisomerase I and gyrase. Prior to any investigations, seek approval for the procedures from your instructor.

## Questions

1) What effects do the enzymes have on the topology of DNA?

2) Can you observe any mechanistic differences between the two classes of enzymes?

3) Which cofactors are necessary for the activity of each enzyme?

## References for Experiment

Keck, M. V. DNA Topology Analysis in the Undergraduate Biochemistry Laboratory. *J. Chem. Educ.*, 77, 2000: 1471-1473.

Tweedie, J. W., and K. M. Stowell. Quantification of DNA by Agarose Gel Electrophoresis and Analysis of the Topoisomers of Plasmid and M13 DNA Following Treatment with a Restriction Endonuclease or DNA Topoisomerase I. *Biochem. And Mol. Biol. Educ.*, 33, 2005: 28-33.

Wang, J. C. DNA Topoisomerases: Why So Many? *J. Biol. Chem.*, 266, 1991: 6659-6662.

# Experiment 25. Effect of Drugs on DNA Topoisomerases

## Purpose of the Experiment

In this experiment, the effect of drugs on DNA topoisomerases will be investigated.

## Prelaboratory Assignment

Predict the effect of the drugs on the appearance of the DNA in the gel.

## Materials

DNA samples, relaxed and supercoiled plasmids
10X Reaction buffer for topoisomerase I (10 mM Tris-Cl, pH 7.5, 100 mM NaCl or KCl,
        1 mM PMSF and 1.0 mM mercaptoethanol)
Topoisomerase I
Sodium dodecyl sulfate (SDS), 10% w/v
Proteinase K (10 mg/mL)
Camptothecin (10 mM in DMSO)
DMSO
Water bath
Electrophoresis apparatus
Agarose
1X TBE buffer (89 mM Tris base, 89 mM boric acid, and 2 mM EDTA, pH 8.0)
DNA electrophoresis loading buffer (30% glycerol, 0.25% (w/v) bromophenol blue, 0.25% (w/v)
        xylene cyanol)
Ethidium bromide or alternative DNA stain

## Safety

Ethidium bromide is a suspected carcinogen. Wear gloves when handling ethidium bromide solutions. Wear UV goggles to protect your eyes when using the UV transilluminator. Dispose of the solutions at the end of the experiment in the proper manner, as indicated by the instructor.

## Experimental Methods

1)  Prepare the mixtures listed in Table 8.6 in microcentrifuge tubes on ice. **Note –** the drug is dissolved in DMSO. DMSO is added to tubes 1 and 3 to ensure that the effects on the topology is due to inhibition of the enzyme by the drug, not by the solvent.

Table 8.6 – Samples for Topoisomerase Studies.

| Reaction Component | Tube | | | | | | |
|---|---|---|---|---|---|---|---|
| | 1 | 2 | 3 | 4 | 5 | 6 | 7 |
| 10 x Reaction Buffer | 2 µL | 2 µL | 2 µL | 2 µL | 2 µL | 2 µL | 2 µL |
| Water | 14 µL | 14 µL | 12 µL | 13.5 µL | 13 µL | 12.5 µL | 12 µL |
| Supercoiled DNA | 2 µL | 2 µL | 2 µL | 2 µL | 2 µL | 2 µL | 2 µL |
| Camptothecin Solution | —— | 2 µL | —— | 0.5 µL | 1 µL | 1.5 µL | 2 µL |
| DMSO | 2 µL | —— | 2 µL | —— | —— | —— | —— |

2) Centrifuge the tubes in a microcentrifuge for a few seconds in order to draw all of the liquid to the bottom of the tube.

3) Add 2 µL of the topoisomerase I to tubes 3-7.

4) Incubate the tubes at 37°C for 30 minutes.

5) Meanwhile, pour a 1% agarose gel in TBE buffer. Based on the manufacturer's information, determine the volume of agarose solution required to create the gel. Measure the appropriate amount of agarose, and place it in an Erlenmeyer flask with the determined volume of TBE buffer. Gently heat the solution over a hotplate, in a microwave, or in an autoclave until the agarose melts and a homogenous solution is created. Avoid boiling or charing the agarose.

6) Cool the solution to 55°C and pour into the gel cast. Remove any bubbles and insert the comb.

7) Allow the gel to harden. Once hardened, remove the gel comb and place the gel into the electrophoresis chamber. Cover the gel with TBE buffer.

8) After incubation, stop the reactions by adding 2 µL of 10% SDS, 2 µL of proteinase K solution, and 4 µL of loading dye. Incubate the samples at 37°C for 15–20 minutes.

9) Slowly load the samples into the wells in the agarose gel using a micropipetter.

10) Close the electrophoresis apparatus. Apply a voltage of 1–10 V/cm until the bromophenol blue has migrated within 1 cm of the end of the gel.

11) Turn off the power, disconnect the leads, and remove the gel from the electrophoresis chamber.

12) Stain the gel by soaking in TBE buffer containing 0.5 µg/mL ethidium bromide for 10 to 30 minutes.

13) Visualize the DNA bands by illuminating the gel with UV light using a transilluminator.

14) Document the gel using photography.

## Questions

1) What different topological forms of DNA are observed on the gel?

2) What effect does the camptothecin have on the enzyme? **Hint:** Think about your answer to Question 1. At what concentration does camptothecin affect the enzyme?

## References for Experiment

Keck, M. V. DNA Topology Analysis in the Undergraduate Biochemistry Laboratory. *J. Chem. Educ.*, 77, 2000: 1471-1473.

Tweedie, J. W., and K. M. Stowell. Quantification of DNA by Agarose Gel Electrophoresis and Analysis of the Topoisomers of Plasmid and M13 DNA Following Treatment with a Restriction Endonuclease or DNA Topoisomerase I. *Biochem. And Mol. Biol. Educ.*, 33, 2005: 28-33.

Wang, J. C. DNA Topoisomerases: Why So Many? *J. Biol. Chem.*, 266, 1991: 6659-6662.

# Experiment 26. Use of Commercially Available Kits for PCR Amplification

## Purpose of the Experiment

In this experiment, DNA will be isolated and amplified using PCR. Commercially available kits (selected by your instructor) will be used to determine either genetic variability among individuals or to determine genetic modification of foods.

PCR reactions are highly dependent on the concentrations of the components used. It is essential that proper pipette techniques are utilized to deliver the small quantities by accurate means. (See Section 1.9.) Also, gloves should be worn to avoid contamination of samples with DNA from fingers.

## Safety

Ethidium bromide is a suspected carcinogen. Wear gloves when handling ethidium bromide solutions. Wear goggles to protect your eyes when using the UV transilluminator. Dispose of the solutions at the end of the experiment in the proper manner, as indicated by the instructor.

## Experimental Methods

Following the instructions provided in the commercial kit, extract the DNA from the biological sample and conduct the PCR amplification.

## Questions

1) What information about your sample did you deduce using PCR? What negative controls could be included in the experiment to confirm your results of the amplification?

2) Is the PCR technique infallible? Explain your answer.

3) Could PCR be conducted without a heat-stable DNA polymerase? If so, what changes in the protocol would be required?